ACTIVATION ANALYSIS
PRINCIPLES AND APPLICATIONS

ACTIVATION ANALYSIS
PRINCIPLES AND APPLICATIONS

PROCEEDINGS OF A NATO ADVANCED STUDY INSTITUTE HELD IN GLASGOW

Edited by

J. M. A. LENIHAN

Western Regional Hospital Board, Glasgow, Scotland

and

S. J. THOMSON

University of Glasgow, Scotland

1965

ACADEMIC PRESS · LONDON AND NEW YORK

ACADEMIC PRESS INC. (LONDON) LTD
BERKELEY SQUARE HOUSE
BERKELEY SQUARE
LONDON, W.1.

U.S. Edition published by

ACADEMIC PRESS INC.
111 FIFTH AVENUE
NEW YORK, NEW YORK 10003

Library of Congress Catalog Card Number: 65-27738

PRINTED IN GREAT BRITAIN BY
SPOTTISWOODE BALLANTYNE AND CO. LTD
LONDON AND COLCHESTER

CONTRIBUTORS

H. J. M. BOWEN, *United Kingdom Atomic Energy Authority, Wantage Research Laboratories, Wantage, Berks, England.* (Chapters 16, 23, 24)

W. BYGRAVE, *High Voltage Servicing Co. Ltd., Eastcote, Middlesex, England.* (Chapter 9)

P. A. CAWSE, *United Kingdom Atomic Energy Authority, Wantage Research Laboratories, Wantage, Berks, England.* (Chapter 24)

G. I. CRAWFORD, *Natural Philosophy Department, the University, Glasgow, Scotland.* (Chapter 17)

D. GIBBONS, *United Kingdom Atomic Energy Authority, Wantage Research Laboratories, Wantage, Berks, England.* (Chapters 4, 5, 10, 13, 15)

V. P. GUINN, *General Atomic Division, General Dynamics Corporation, San Diego, California, U.S.A.* (Chapters 8, 11, 19, 20)

J. M. A. LENIHAN, *Regional Physics Department, West Graham Street, Glasgow, Scotland.* (Chapters 1, 3, 18)

G. S. NIXON, *University of Glasgow Dental Hospital and School, Renfrew Street, Glasgow, Scotland.* (Chapter 21)

HAMILTON SMITH, *Department of Forensic Medicine, The University, Glasgow, Scotland.* (Chapter 22)

S. J. THOMSON, *Chemistry Department, The University, Glasgow, Scotland.* (Chapter 12)

R. E. WAINERDI, *Texas A. & M. University, College of Engineering, College Station, Texas, U.S.A.* (Chapters 4, 6, 7)

J. E. WHITLEY, *Scottish Research Reactor Centre, East Kilbride, Glasgow, Scotland.* (Chapter 25)

H. W. WILSON, *Scottish Research Reactor Centre, East Kilbride, Glasgow, Scotland.* (Chapters 2, 14)

Experimental Section

K. C. CAMPBELL, D. GIBBONS, I. M. HOODLESS, A. HOWIE, J. M. A. LENIHAN, B. McANDREW, HAMILTON SMITH, S. J. THOMSON, R. E. WAINERDI, J. E. WHITLEY, H. WILSON.

FOREWORD

Activation analysis is an experimental technique applicable to most of the chemical elements; for more than half of them it provides a sensitivity unequalled by any other method. With the growing availability of research reactors, discharge-tube neutron sources and sophisticated electronic equipment for radioactive assay, the potential uses of activation analysis in science, medicine and industry are becoming more widely appreciated.

The NATO Advanced Study Institute held in Glasgow in August, 1964, was organized to provide a comprehensive review of the principles and applications of activation analysis for a class of twenty students, all at post-doctoral level of qualification or experience. The students were drawn from many countries and many disciplines. Their curriculum had four main parts. First came a series of lectures on the scientific and technical principles of activation analysis, with a programme of illustrative practical exercises at the Scottish Research Reactor Centre, East Kilbride.

The class then migrated to the University of Glasgow where a number of laboratories had been equipped for more intensive practical study. A large exhibition of equipment, provided by European and American manufacturers was mounted at this time. Students were able to perform many experiments using the newest apparatus under realistic conditions and were able to study a wide range of commercially available apparatus, including neutron generator systems in full operation.

Finally a series of colloquia were given at the Regional Physics Department of the Western Regional Hospital Board, Glasgow. Here, as in earlier sections of the course, visiting experts from several British and American laboratories contributed to teaching and discussion.

This book contains a record of the proceedings of the Advanced Study Institute. Contributors were not asked to provide written texts, but reports of the lectures, prepared by a team of graduate and undergraduate assistants, were approved and, in some instances, substantially extended by the original speakers. The detailed instructions for the laboratory exercises are also included in the text, which is offered as a survey of the present state of activation analysis, based on the experience of people busily engaged in relevant basic research or in the pursuit of practical applications.

We are grateful to the contributors for their help in preparing the lecture texts. In the belief that the distinctively spontaneous and informal flavour of the lectures and colloquia is worth preserving, we have not been too scrupulous in eliminating overlapping or duplication. It is a pleasure to acknowledge the help provided by our team of recorders, Mr. M. G. Albrow, Mr. B. McAndrew and Mr. J. F. Wright.

July 1965

J. M. A. Lenihan

S. J. Thomson

CONTENTS

ACTIVATION ANALYSIS: PRINCIPLES AND APPLICATIONS

EXPERIMENTAL SECTION

ACTIVATION ANALYSIS
PRINCIPLES AND APPLICATIONS

Chapter I

RADIOACTIVITY

J. M. A. LENIHAN

Regional Physics Department, 9–13 West Graham Street, Glasgow, C.4, Scotland

The existence of the material universe, and the geochemical and cosmological evolutionary processes taking place within it, depend to a large extent upon the long-term stability of the atoms of which it is composed. It is ironic that most of the nuclear species are unstable, and from this point of view one may say that the normal state of matter is unstable. Of the known nuclides approximately 1500 are radioactive, while only 300 are stable; these stable isotopes are, however, far more abundant because of their permanence.

The process of radioactivity is the means by which an atomic nucleus increases its stability by emission of radiation in the form of electrons or α-particles. This can be understood in terms of a stability diagram (Fig. 1) in which the number, N, of neutrons in a nucleus is plotted against the number, Z, of protons. It will be seen that the stable nuclei lie approximately along a line which may be called the line of stability, and that the ratio $Z:N$ can therefore only take a limited range of values. In radioactive decay the values of both N and Z change, either by the conversion of a neutron into a proton, or by the reverse process in which a proton is converted into a neutron. These processes may be written

$$n \longrightarrow p^{+}+e^{-}+\bar{\nu} \qquad (1)$$

$$p^{+} \longrightarrow n+e^{+}+\nu \qquad (2)$$

In these transformations either a positive or a negative electron is emitted by the nucleus, and this radiation is the β-radiation. A neutrino (ν) or anti-neutrino ($\bar{\nu}$) is also emitted; these are particles with neither mass nor charge which allow energy and momentum to be conserved in the nuclear reaction. They are not readily detectable, and need not concern us here. Let us consider the transformation process (1) above, in which a neutron is converted into a proton. The atomic number Z (which equals the number of protons) increases to $(Z+1)$, while the neutron number decreases to $(N-1)$. The radioactive nuclide thus moves one place diagonally down to the right in the stability diagram by emitting a negative electron. Unstable nuclei, on the upper left of the stability line in Fig. 1, are therefore expected to be electron emitters. By a similar argument it will be seen that unstable nuclei on the lower right of the

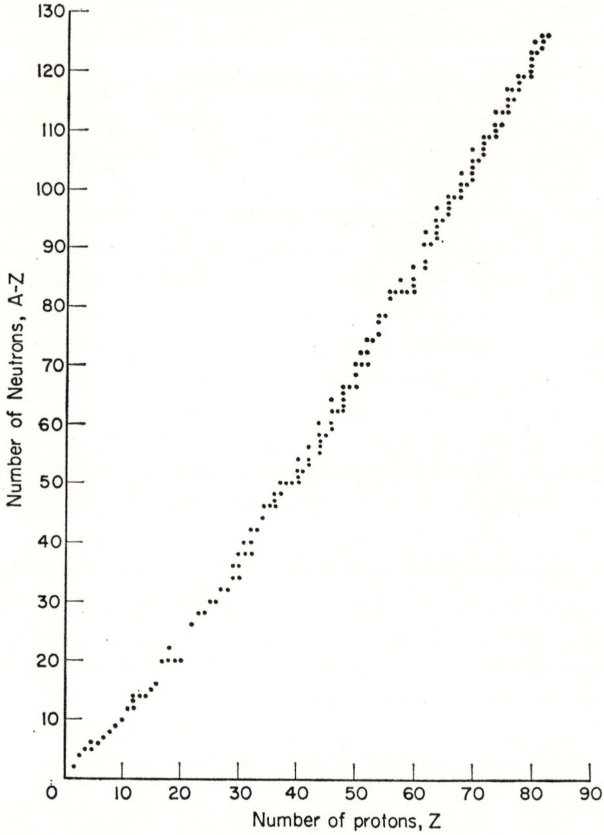

FIG. 1. Stability diagram. Plot of the number of neutrons in the nucleus, N against the number of protons, Z.

stability line probably emit positive electrons, or positrons, when they decay. As the distance of the unstable nucleus from the stability line increases, the half-life of the nuclides tends to become shorter, and detection of the nuclide becomes increasingly difficult.

After a nuclear transformation has taken place by β-decay, the daughter atom may itself be unstable. This nucleus may have a surplus of energy, i.e. it may be excited, and this energy is usually emitted in the form of γ-rays (although other ways of releasing the energy are important and will be discussed below). In the nuclear reaction,

$$^{32}\mathrm{P} \longrightarrow {}^{32}\mathrm{S} + e^- \tag{3}$$

the sulphur nucleus is formed in the ground state and is stable. However, after sodium decay,

$$^{24}\mathrm{Na} \longrightarrow {}^{24}\mathrm{Mg}^* + e^- \tag{4}$$

$$^{24}\mathrm{Mg}^* \longrightarrow {}^{24}\mathrm{Mg} + \gamma \tag{5}$$

the magnesium daughter nucleus is left in an excited state—denoted by an asterisk—from which it decays with the emission of a γ-ray. In this case the γ-ray has a well defined energy and gives rise to a well defined spectral line. It is possible for the γ-ray emission to take place in more than one stage via intermediate excited states of the daughter nucleus, and in this case two or more spectral lines will be present in the γ-ray spectrum. It should be remembered that the energies of the γ-rays are characteristic of the daughter nucleus, and not of the parent radioactive nucleus; in fact, they enable the daughter nucleus to be identified.

An excited nucleus may also release its energy by emitting an extra-nuclear electron from the atom, rather than a γ-ray. This ejected electron will have very nearly the same energy that an emitted γ-ray would have had; the difference is equal to the energy necessary to separate the negative electron from the positive ion that remains, i.e. it is the ionization energy of that electron. This process is called internal conversion.

A β-radioactive nucleus may also decay in modes other than emission of an electron from the nucleus, for the ratio $Z:N$ may also be altered by the absorption of an electron into the nucleus from the K-shell of the atom. This process is called K-capture, and it competes with positron decay. The vacancy left in the K-shell after electron capture is filled by electrons falling in from the outer shells, and K-capture may be detected by the X-radiation these electrons emit as they change their energy levels.

These are the main forms of radioactive decay, apart from α-particle emission which has very few applications in activation analysis. There are other more subtle forms of decay but a description of these would also be out of place here.

There are several types of instrument with which the radiation from radioactive substances can be detected and measured. Geiger counters are normally used for the detection of β-particles while γ-rays are detected by a scintillation counter. The γ-ray cannot be detected directly, but only because it transfers energy to electrons, by either the photo-electric effect, the Compton effect or (for high energy γ-rays) the pair production process.

In investigations of radioactive decay there are three important features by means of which the process can be identified. The emitted radiation is, of course, important; in the cases in which we are concerned this may be a positive or negative β-particle or a γ-ray. The second feature is the rate of the exponential decay of the radioactivity, as measured by the half-life, $t_{1/2}$. This is the time in which half of the radioactive nuclei in any sample will decay. Some nuclides have a half, life of a small fraction of a second, but ^{238}U has a half-life of several thousand million years, which is comparable to the age of the earth. It is this fact that accounts for the presence of naturally occurring radioactive materials. The third important observable is the energy involved in the transformation, and in particular the energies of any γ-rays emitted, which generally lie between a few keV and a few MeV (million electron volts).

From observations of these three features and the use of data published in tables, it is possible to discover which isotopes are responsible for the radioactivity of a particular sample. This broadly constitutes the practice of γ-ray spectroscopy, which is somewhat analogous to optical spectroscopy, though the resolving power attainable is very much lower.

In order to place the subject of radioactivity on a sound quantitative basis, various units of measurement have been proposed. None of these has been found wholly satisfactory. Originally radium was the only material used in radioactive studies, and a Curie was defined as the radioactivity associated with 1 g of radium. When other active substances came into use, this unit was redefined as the amount of a radioactive material in which $3{\cdot}7 \times 10^{10}$ atoms disintegrate per second (this is approximately equal to, but more precisely defined, than the old Curie).

The radioactivity of a specimen can be measured by direct counting, but it is difficult in practice to record every disintegration. Another method, suitable for pure β-minus emitters, which is rather less precise, is to measure the total energy released, using a calorimeter (all of the energy can ultimately be dissipated as heat). If the energy per disintegration is known from the decay scheme, the number of disintegrations can then be calculated. However, for most purposes in activation analysis, it is sufficient to compare the unknown radioactivity with that of a standard which has been treated identically both physically and chemically.

The key to the widespread applications of activation analysis is the very great sensitivity of detection of radioactivity which is possible. The lower limit for most analytical techniques is about 10^{15} atoms. Activation analysis can detect samples of only 10^9 atoms in certain favourable cases, which is a million-fold improvement in sensitivity. It is rare to find such a large technical advance occurring in one stage; for this reason, many of the applications which are feasible have not yet been exploited fully. The potential of activation analysis is such that it may assume an importance even greater than nuclear power.

Chapter 2

NUCLEAR REACTIONS

H. W. WILSON

Scottish Research Reactor Centre,
East Kilbride, Glasgow, Scotland

In normal chemical reactions the molecular arrangements are changed with the emission or absorption of energy in the form of heat or light. For example in the chemical reaction $C + O_2 \rightarrow CO_2$, approximately 4·1 eV energy is released in each reaction (an electron volt, eV, is the energy acquired by an electron in falling through a potential difference of 1 V; 1000 eV = 1 keV and 1000 keV = 1 MeV). In a nuclear reaction the structure of the atomic nucleus itself undergoes a change, and the energies involved are far greater. As an example, a nitrogen atom nucleus may react with a neutron to form a carbon nucleus and a proton, with 620 keV of energy. This is written:

$$^{14}_{7}N + ^{1}_{0}n \longrightarrow ^{14}_{6}C + ^{1}_{1}H + Q$$

In this formula the subscripts denote the atomic number or charge, Z, of the particle and the superscripts denote the atomic weight or mass number, A. Both these numbers must balance on each side of the equation. Q is the energy released per reaction (+620 keV in this case) and may be positive or negative. In the former case energy is released and the reaction is said to be exothermic, while in the latter case (Q negative) the reaction is endothermic and energy is absorbed in the reaction. The energies involved per reaction are often a million or more times greater than those of chemical reactions. The value of Q may be calculated from tables showing the masses of the nuclei and particles involved in the reaction. In the reaction cited above, a proton and a ^{14}C nucleus together have a mass less than the combined masses of the neutron and the ^{14}N nucleus, by an amount Δm. This mass difference appears in the form of energy, Q, according to Einstein's relation, $E = mc^2$, or in this case $Q = \Delta m.c^2$ where c is the velocity of light. Nuclear reactions are usually written in a shorthand notation in which the above reaction becomes

$$^{14}_{7}N(n,p)^{14}_{6}C$$

Other abbreviations used are p (proton) for ${}^1_1$H; d (deuteron) for ${}^2_1$H; t (triton) for ${}^3_1$H; and α (alpha) for ${}^4_2$He.

The first nuclear reaction to be observed in the laboratory, by Rutherford in 1919, was of the (α,p) type. α-Particles from a radioactive source reacted with ${}^{14}_7$N nuclei in the air and gave rise to protons, changing the nitrogen nuclei into oxygen in the process.

The first artificial nuclear disintegration was achieved by Cockcroft and Walton who, in 1932, bombarded lithium with artificially accelerated protons. A pair of α-particles are produced by the reaction

$$ {}^7_3\text{Li}(p,\alpha){}^4_2\text{He} $$

and were observed as scintillations on a zinc sulphide screen.

When nuclear reactions are caused by charged particles, protons or α-particles, the bombarding missile must have enough energy to overcome the electrostatic repulsion of the positively charged nucleus. This repulsion may be said to produce a "coulomb barrier" around each nucleus which can only be penetrated by charged particles of sufficient energy. On the other hand, neutrons have no charge and even very low energy neutrons can enter a nucleus and cause a reaction.

Activation analysis is very largely concerned with neutron reactions. One of these is the neutron-fission reaction which takes place in the core of a nuclear reactor and enables the reactor to be used as a neutron source. In this type of reaction a neutron enters a heavy nucleus, such as ^{235}U, which becomes unstable, breaking up into two approximately equal halves and giving off a few surplus neutrons (and energy). Because, on average, each reaction produces more neutrons than the single one necessary to start the reaction, these neutrons can initiate other fissions and a chain reaction can occur. The neutrons can also be used to bombard samples and make them radioactive (for analysis or other purposes) by the neutron capture reaction, (n,γ). For example ^{23}Na absorbs a slow neutron to form an excited ^{24}Na nucleus, which then releases its surplus energy in the form of a γ-ray. ^{24}Na is radioactive decaying by beta and gamma emission and can thus be detected by a counter. Neutron capture reactions are all exothermic.

In activation analysis the trace elements are usually made radioactive by a (n,γ) reaction and it is important to consider possible sources of neutrons. The earliest neutron source (used by Chadwick when he discovered the neutron in 1932), was a combination of an α-radioactive material and the light element beryllium. The α-particles interact with beryllium to form carbon nuclei and neutrons according to the reaction,

$$ {}^9_4\text{Be}(\alpha,n){}^{12}_6\text{C} $$

A mixture of radium and beryllium is still used as a neutron source for some purposes, but the neutron flux is very low compared with other sources. A modern neutron generator usually has a target bombarded by accelerated deuterons. Fast neutrons are produced by the reactions.

$$ {}^{2}_{1}\mathrm{H}(d,n)\,{}^{3}_{2}\mathrm{He} + 17{\cdot}6\ \mathrm{MeV} $$

$$ {}^{3}_{1}\mathrm{H}(d,n)\,{}^{4}_{2}\mathrm{He} + 3{\cdot}25\ \mathrm{MeV} $$

Before they can produce a neutron capture reaction, fast neutrons usually have to be slowed down or thermalized in paraffin or another hydrogenous material, because slow neutrons have a far greater chance of being captured by target nuclei. High energy neutrons often cause reactions other than neutron capture such as the $(n,2n)$, (n,p), and

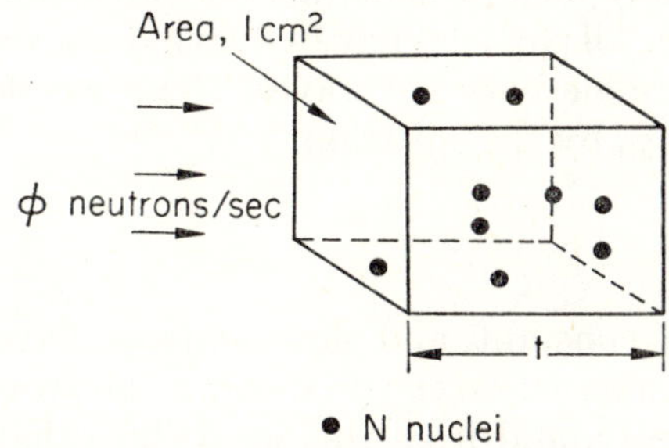

FIG. 2. The derivation of cross-sections.

(n,α) reactions. ^{14}N can absorb a neutron and then release two neutrons to become ^{13}N, which is a positron emitter with a half-life of 12·3 min. ^{16}O can emit a proton after irradiation with fast neutrons, and it then transforms into ^{16}N which has a 7·4 sec half-life.

The term "cross-section" is used in the description of nuclear reactions; it measures the probability of a reaction taking place. In Fig. 2 a flux of ϕ neutrons per sec is incident upon one square centimetre of material. The material has a thickness t and contains N nuclei. The probability, P, of a reaction taking place in 1 sec is proportional to N and to ϕ, and we may call the proportionality constant σ. Thus,

$$ P = \sigma N \phi $$

When dimensions are equated for this equation, it will be seen that σ represents an area, and can be considered as equal to the area presented by each nucleus to the neutron beam.

The cross section is measured in units of area called barns, where

$$1 \text{ barn} \equiv 10^{-24} \text{ cm}^2$$

The radius of the nucleus is given approximately by the equation,

$$r = r_0 A^{1/3}$$

where r_0 is a constant, of magnitude $1{\cdot}2\text{–}1{\cdot}5 \times 10^{-13}$ cm, depending on how it is defined, and A is the nuclear mass number.

For nuclei of medium mass number, the cross section,

$$\sigma = \pi r^2$$

comes out to be of the order of a few barns and this is why the unit is so chosen.

This geometrical discussion is highly simplified, for a nucleus cannot be regarded as a solid sphere, and neither does an incident particle always behave as a particle. All particles have a dual nature, sometimes behaving as "particles" and sometimes as "waves". The wavelength, λ, is related to the momentum, p, by the equation,

$$\lambda = \frac{h}{p} = \frac{h}{mv}$$

where h is Planck's constant, and slow neutrons have therefore a large wavelength. The cross-section can be shown to be proportional to $\lambda^2 E^{1/2}$, where E is the energy, and from this we deduce that the cross-section for neutron reactions increases as the velocity decreases according to the law,

$$\sigma \propto 1/v$$

where v is the neutron velocity. On the particle picture one can picture this crudely as meaning that the probability of a reaction occurring is proportional to the amount of time that the neutron spends in the vicinity of the nucleus. The absolute value of the (n,γ) cross-section at thermal energies varies over a very wide range of values from isotope to isotope, e.g. it is $5{\cdot}7 \times 10^{-4}$ barns for $^2_1$H and $2{\cdot}7 \times 10^6$ barns for ^{135}Xe.

Further Reading

1. Friedlander, G., Kennedy, J. W. and Miller, J. M. (1965). *Nuclear and Radiochemistry*, 2nd edn. London: Wiley.
2. Kaplan, I. (1963). *Nuclear Physics*, Chapts. 11, 16, 18 and 19. Mass.: Addison-Wesley.

Chapter 3

SCINTILLATION COUNTING

J. M. A. LENIHAN

Regional Physics Department, 9–13 West Graham Street, Glasgow, C.4, Scotland

The first radiation detector used to count relative rates of disintegration was the "spinthariscope", a thin disc of fluorescent material placed in front of the eye, the final detector. This method was employed by Rutherford and his associates. The first instrument which could properly be called a scintillation counter was devised in 1944 by Curran and Baker, who employed a thin layer of zinc sulphide coupled to a photomultiplier. The disadvantage of this method was that the ZnS layer had to be thin enough to be transparent to the fluorescent emission, and at the same time thick enough to absorb a considerable fraction of the incident γ-radiation.

With ZnS these two requirements could not easily be met, but with NaI, as used later, one has a material which is sufficiently transparent in the visible range to permit the use of a thick detector, thus absorbing a greater fraction of the incident radiation. The scintillation counter initiated an important advance in radiation detection and analysis.

Phosphor Materials

Inorganic materials such as NaI and CsI have a high fluorescent efficiency and high absorption (related to Z) for γ-radiation but generally have a fairly slow decay. In organic materials (e.g. anthracene) with a high proportion of H, C and O atoms the average Z value is about 7, giving a lower absorption cross-section and hence a smaller light output. This is offset, however, by a faster decay. Liquid phosphors, consisting of fluorescent material dissolved in compounds such as toluene, are useful in such fields as low-energy β decay. For some purposes one can employ the inexpensive plastic phosphors, which are comparable to liquid scintillators, in that they are manufactured by polymerization of solutions of scintillating materials in a styrene or vinyltoluene solvent.

The scheme shown in Fig. 3 illustrates a typical scintillation counting arrangement in diagrammatic form.

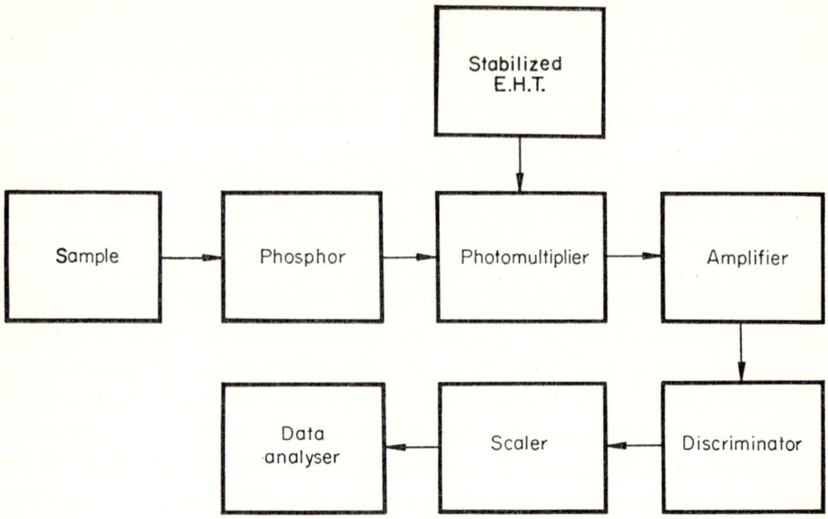

FIG. 3. Typical scintillation-counting arrangement.

GAMMA-RAYS: INTERACTION WITH MATTER

When a γ-ray quantum is incident on a material its energy can be transferred in any or all of three ways. The cross-sections for these events depend on the energy of the γ-ray, as shown in Fig. 4.

Photoelectric absorption, important for absorbers of high atomic number or radiation of relatively low energy, involves the ejection of one

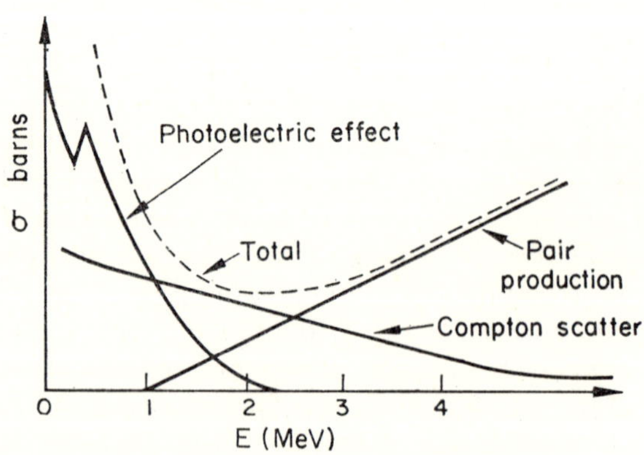

FIG. 4. Cross-sections for γ-absorption.

of the orbital electrons from an atom. The outer electrons are more easily removed than those closer to the nucleus.

The cross section for photoelectric absorption decreases with increasing γ-ray energy. The peaks in this curve correspond to the energy levels of the inner electrons, the probability of absorption being greater when the γ photon has exactly (or slightly more than) the energy required to eject the electron from the atom.

Compton scattering is the transfer of part of the energy of a quantum to an electron. The scattered quantum has a longer wavelength, the difference in initial and final energies being dependent on the angle of scattering of the electron relative to the initial direction of the quantum.

The remaining possible interaction is that of pair production, in which the quantum vanishes completely and its energy appears in the form of β^+- and β^--particles. The threshold for this phenomenon is at 1·02 MeV, since the rest mass energy of each particle is 0·51 MeV. If the energy of the quantum exceeds 1·02 MeV, the excess energy appears as the kinetic energy of the β-particles. The probability for pair production increases with energy. The reverse effect, electron-positron annihilation, is also possible. Here the two particles collide and their energy and mass are converted into electromagnetic radiation. In this case, however, instead of one quantum being formed, the demands of conservation of energy and momentum necessitate the creation of two quanta, moving in opposite directions, each of energy 0·51 MeV.

As the result of all these processes, the signal obtained from the photomultiplier may be quite complicated. Consider, for example, a monoenergetic γ-emitter, such as ^{28}Al. The disintegration of this species is relatively simple, yielding from each atom a single γ-ray quantum of energy 1·78 MeV. Ideally we should hope that each quantum would result in a single light pulse, so that all of the electrical pulses examined by the discriminator (or pulse-height analyser) would be identical and would correspond to an initial quantum energy of 1·78 MeV.

Unfortunately this desirable result is never obtained with scintillation counters of the kind that we are now considering (though newer devices, based on semi-conducting materials, come nearer to the ideal performance). To understand and interpret the information obtained from a scintillation counter it is necessary first to appreciate that the essential effect involved is the transfer of energy to electrons. A quantum which passes right through the phosphor unaltered in energy will convey no information to the photomultiplier or the subsequent electronic devices. The fluorescent process is initiated by the transfer of energy from the quantum to an electron. It is useful to consider how this transfer takes place in the various absorption processes already mentioned.

1. PHOTOELECTRIC ABSORPTION

Here the major part of the energy of the quantum may be transferred to the electron. A certain amount of energy is needed to separate the electron from the electrostatic attraction of the nucleus; for light elements (such as H, C, N and O) this is usually only a few eV and even for heavier elements the energy loss is very small unless an inner shell electron is detached, when the energy required may reach several keV. Generally, then, the electron will take virtually all of the quantum energy. If photoelectric absorption was the only process involved, we should expect exposure of a phosphor to the γ-rays from ^{28}Al to give a sharply defined energy spectrum (Fig. 5).

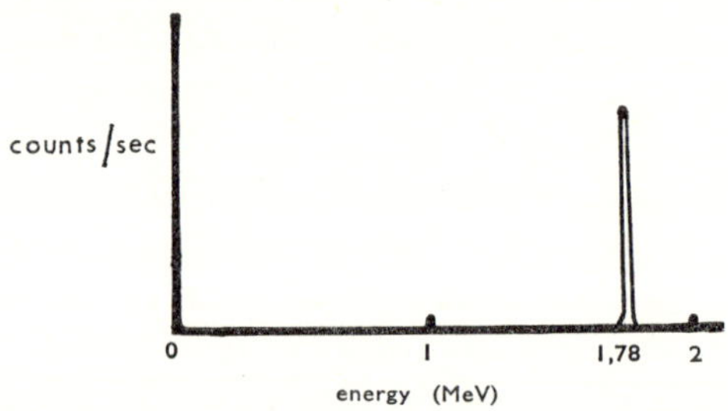

FIG. 5. Idealized γ-ray spectrum of ^{28}Al assuming only photoelectric absorption.

2. COMPTON ABSORPTION

It is not possible to obtain photoelectric absorption alone. The accompanying Compton processes add to the complexity of the eventual energy spectrum. The initial energy of the quantum is divided between the electron and the scattered quantum in a way that depends on the angle of scattering and on the energy of the original quantum. For a given initial quantum energy, there is an upper limit to the amount of energy that can be transferred to the electron. The maximum energy transfer occurs when the quantum is scattered through an angle of 180°. For a high energy quantum, the maximum energy acquired by the electron in the Compton process will be approximately 0·25 MeV less than the initial quantum energy. For lower initial energies, the gap (which corresponds, of course, to the energy removed by the scattered quantum) will be somewhat less. For angles of scattering less than 180°, the energy transferred to the electron will be less than the maximum. In practice, most of the quanta are scattered through angles less than 180° and the energy distribution of the output pulses delivered from the photomultiplier has the appearance shown in Fig. 6. The maximum electron energy is known as the Compton edge and the rest of the spectrum is referred to as the Compton continuum. For the ^{28}Al γ-radiation, the Compton edge is at 1·56 MeV.

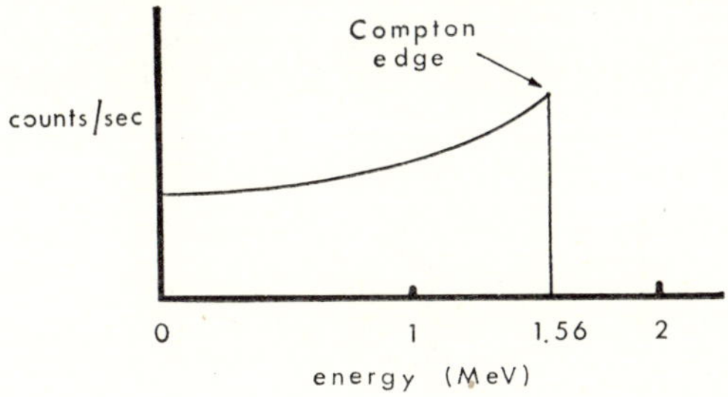

FIG. 6. Idealized energy spectrum of electrons released in Compton scattering.

Subsequent photoelectric absorption of the quanta scattered through 180° gives rise to a back-scatter peak at an energy which is never more than 0·255 MeV and is generally in the region of 0·2 MeV, depending on the energy of the original γ-ray quantum.

3. PAIR PRODUCTION

The 1·78 MeV γ-ray quantum from ^{28}Al may lead to pair production and—after a very short time—annihilation of the positron-electron pair with release of two 0·51 MeV quanta. If both of these quanta escape from the phosphor, the maximum amount of energy that can be transferred

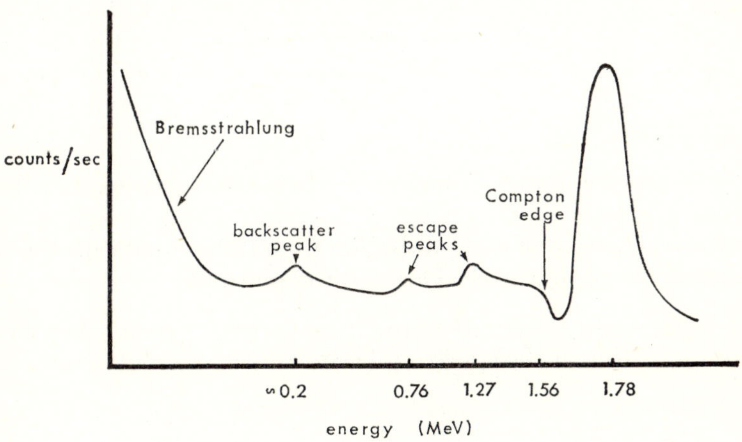

FIG. 7. Typical γ-ray spectrum from ^{28}Al.

to electrons in the phosphor will be 0·76 MeV. If only one of the annihilation quanta escapes, the residual energy is 1·27 MeV. If both of the 0·51 MeV quanta are absorbed inside the phosphor, by photoelectric effect, a single photo-peak of energy 1·78 MeV can be expected. In general, we may expect to find two of the "escape peaks" in the energy spectrum—one at 0·51 MeV below the energy of the quantum involved and another at 1·02 MeV below.

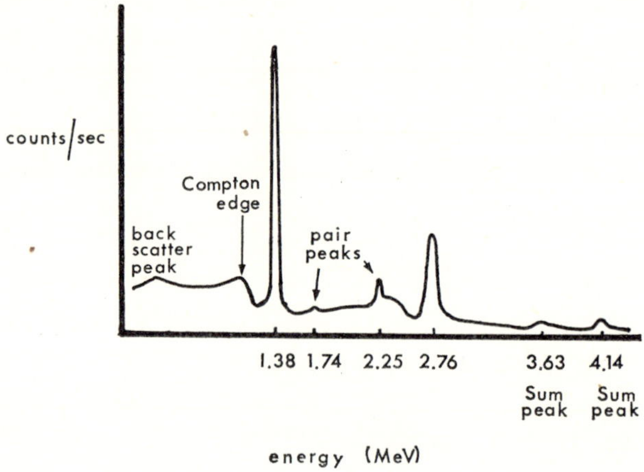

FIG. 8. Typical γ-ray spectrum from ^{24}Na. Sum peaks result from simultaneous photoelectric absorption of two γ-ray quanta; escape peaks may be associated with the sum peaks.

The complete γ-ray spectrum from a sample of ^{28}Al has the general appearance shown in Fig. 7, that from ^{24}Na in Fig. 8. The detailed shape of the spectrum depends on the size and composition of the phosphor as well as on the multiplier and associated electronic equipment. In practice, further complication is introduced by a number of minor effects, such as

(*a*) bremsstrahlung (X-radiation produced as electrons slow down in the phosphor), and
(*b*) the interaction of the γ-radiation from the sample with the lead or other shielding material around the phosphor.

When, as normally occurs, the sample emits two or more γ-ray energies, the interpretation of the spectrum becomes even more difficult.

Chapter 4

DETECTION OF RADIOACTIVITY

D. GIBBONS

*United Kingdom Atomic Energy Authority,
Wantage Research Laboratory, Wantage, Berks., England*

R. E. WAINERDI

*Texas A. & M. University, College of Engineering,
College Station, Texas, U.S.A.*

The principal forms of radiation which are of interest in activation analysis are as follows:

(*a*) β-particles and positrons, having energies of 0·5–3 MeV;
(*b*) X- and γ-rays in the keV and 0·1–3 MeV ranges respectively;
(*c*) α-particles (limited number of applications);
(*d*) neutrons with energies of up to around 14 MeV.

We propose to discuss the detection of these forms of radiation and we shall begin with methods which depend upon the ionization of gases.

Gas Ionization Methods

The elementary form of detector under this heading is the Ionization Chamber. This consists of a pair of electrodes, one at a high and the other at earth potential. The gas between them is kept at a constant pressure. Ion pairs, formed by interaction of the incident radiation with the gas atoms, are collected by the electrodes, and the resulting current produces a signal which is amplified and recorded by the associated electronic equipment.

With this detector the probability of direct ionization is high for β^+- and β^--radiation but almost zero for γ-rays, although with the latter there is a small chance of ionization occurring on the walls of the chamber. When the electrons and ions have been collected by the electrodes, the chamber is then ready to detect the next ionizing event.

If one plots the number of ion pairs produced by an event, against the applied voltage, one obtains a curve of the type shown in Fig. 9.

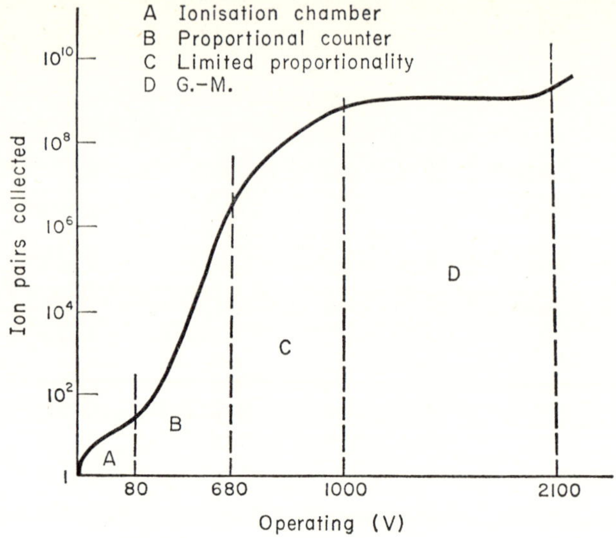

FIG. 9. Detector operating regions.

The curve can be divided into several regions and the shape explained as follows:

(*a*) In the ion chamber region of the curve, at low applied voltage, there is no multiplication of ions, and thus the signal obtained is usually very low. In general, this part of the curve is not of interest in activation analysis, but chambers can be used in this region for counting high activities.

(*b*) In this, the proportional region, the electrons produced by an event are sufficiently accelerated by the electric field to enable them to cause secondary ionization, this process building up until a shower of electrons is collected by the anode. In this way, the gas amplification increases the magnitude of the signal to a convenient level, although further electronic amplification is still necessary.

(*c*) This region is one of limited proportionality in which the increased field is partly offset by the space charge limiting effect of the particles produced by ionization.

(*d*) This is the well-known "Geiger Plateau". At this range of voltage the space charge limiting effect takes over, almost cancelling any increase in voltage. Any ionizing event occurring in this range will give a signal of a fixed magnitude.

(*e*) In the region above the Geiger plateau a condition of continuous breakdown exists.

It should be noted that although the Geiger plateau is suitable for non-proportional counting, the large number of ion pairs produced by an event renders the chamber ineffective in detecting any other event which occurs during the formation of the electron shower. The time interval between the original event and the complete collection of the current is known as the "Dead Time". By plotting a curve of the height of the output pulse obtained from a *second* event as a function of the time elapsed since the beginning of the *first* event, the form shown in Fig. 10 would be obtained.

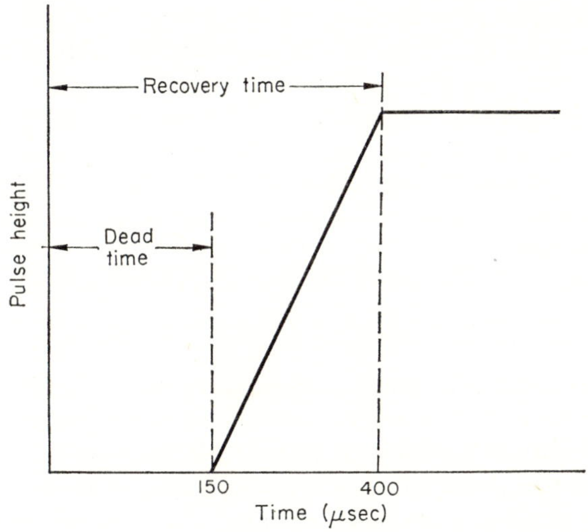

FIG. 10. Recovery-time and dead-time for a Geiger–Muller counter.

During the dead time the output from a second event would be zero. At a time interval greater than about 150 μsec, the output pulse would be dependent on the elapsed time, up to a point at which the output for a second event would be the same as that for the first event. The total time interval from the first event to the point of full recovery is called the recovery time.

Since the dead time depends on many factors, it is good practice to control the total inert period by arranging that the associated electronics have a larger, but constant paralysis time.

The advantages of the Geiger counter are those of uniform pulse height and simplified instrumentation. Counters operating in the proportional region have the ability to discriminate between different types of radiation of different energies. They also have a much shorter dead time, and since, in general, more information is contained in a proportional count, these detectors have a more promising future.

Scintillation Counting

Scintillation counters incorporate a phosphor with the ability to convert some fraction of the energy of an incident quantum into a signal, a photomultiplier tube, and some sort of counting and discriminating facility.

On passage through the phosphor, the γ-ray loses energy to the electrons of the phosphor by the photoelectric effect, Compton scattering or pair production ($E > 1{\cdot}02$ MeV) processes. The excited electrons are liberated into the phosphor lattice and are soon bound to another atom. The energy level, or site, vacated by an electron may be filled by another electron, causing a quantum of light to be emitted as the electron changes from a higher, loosely-bound state, to a lower tightly-bound state. The energy difference between the two states corresponds to the energy of the light quantum.

The phosphor has the photocathode of a photomultiplier on one side of it so that electrons will be emitted for light quanta which "strike" this face.

These electrons will be multiplied by secondary emission from a series of dynodes held at increasing potentials, the final current being transformed to a millivolt signal by means of a capacitor.

In general, the proportion of signals arising from the photoelectric effect in the phosphor is about 15–20%, the remainder being mainly from Compton scattering, which constitutes a sharing of the quantum energy by several electrons.

A disadvantage arising from the use of the photomultiplier is that statistical variations in electron output limit the resolution to about 6% for monochromatic input.

Anti-coincidence Compton Reduction

The output pulse from a scintillation counter is of the form shown in Fig. 11, the rise time of the leading edge being characteristic of the phosphor material, the main factor being the electronic relaxation time for phosphorescence.

For a NaI phosphor, t is $0{\cdot}25\ \mu$sec.

By placing the active sample in close contact with a NaI crystal and surrounding this on three sides by a CsI phosphor (Fig. 12), the Compton-scattered γ-rays will produce phosphorescent emission from both NaI and CsI, providing that the radiation scattered in NaI does not remain in this material until its energy is completely absorbed. In this way a photocathode covering the end of the CsI phosphor will probably detect simultaneously a composite phosphorescence from NaI and CsI.

If the γ-ray suffers one Compton collision in the NaI phosphor, it may lose, say, half its energy before entering the CsI phosphor, where it will

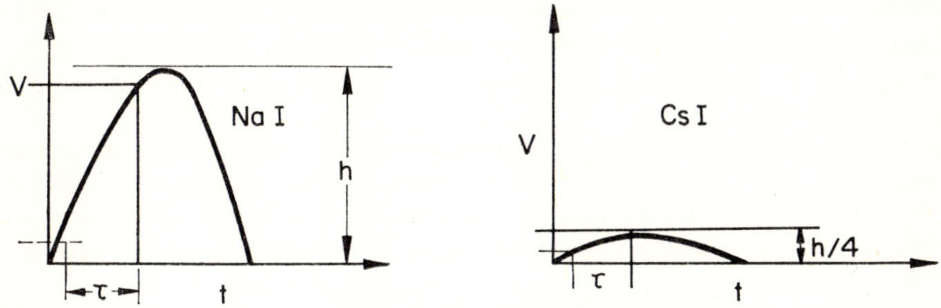

FIG. 11. Pulse shapes from crystal detectors.

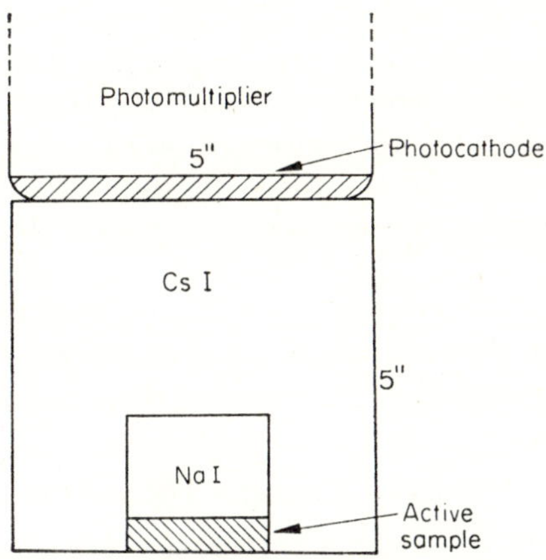

FIG. 12. Combined crystal detector.

lose the remaining half. The output height from CsI is one-quarter of that from NaI for quanta of equal energy, so that the division of output in the case already mentioned is half from NaI and $\frac{1}{2} \times \frac{1}{4} = \frac{1}{8}$ from CsI. In addition, the rise time of a pulse from CsI is longer than the 0·25 μsec for NaI, thus the rise time of the composite pulse is slightly slower than that of the single NaI pulse. Now if the pulse counting circuit incorporates a differentiating unit, the composite pulses can be rejected, leaving the single photoelectric pulses unaltered.

In this way about 80% of Compton scattered radiation can be rejected, rendering the final spectrum considerably better with respect to resolution, and reducing the amount of undesired information.

Solid-State Detectors

In theory solid-state detectors should be extremely useful in many applications because of their size and durability, but in practice the expected reliability and the potential resolution of $\frac{1}{4}$–$\frac{1}{2}$% has not yet been attained.

Materials which come into the semiconductor class, contain as charge carriers "holes" or vacancies in addition to the electrons. A "hole" is simply the vacant site in a crystal lattice which should contain an electron, and because of the surrounding ions it has associated with it a net positive charge. It can be seen that if an electron "falls" into this site, the positive charge will vanish and can be thought to have moved to the vacant site previously occupied by the electron. Thus holes can be regarded as charge carriers, moving in the opposite direction to the electrons.

The first solid-state detector was a small block of semiconducting material with a voltage applied across it. Radiation incident on it should cause ionization and subsequent formation of free electrons and holes. The applied field then draws these in opposite directions, and the resultant current flow constitutes a signal. This system worked, but only for the first signal.

In fact, this type of detector has not proved to be very useful, the main disadvantage being thermal noise, current noise and "flicker" noise, which was dependent on the frequency of the signal.

Surface Barrier Detector

In the silicon surface barrier detector, a *p*-type surface is formed on high purity *n*-type silicon producing a *p*–*n* junction. The application of a reverse bias depletes the junction region of charge carriers and establishes a field across this depletion region. Ionizing radiations, incident on the silicon, produce electron-hole pairs, where the energy required per pair is 3·5 eV, independent of the nature of the radiation. An incident particle, losing E MeV in the depletion region, will produce

$$N = \frac{E \times 10^6}{3{\cdot}5}$$

electron-hole pairs, corresponding to charge $Q_D = 1{\cdot}6N \times 10^{-19}$ coulombs. The charge carriers move in the depletion region field, and, if they traverse a fraction, f, of the depletion depth, w, the charge induced on the terminals is $Q_c = fQ_D$. If carrier losses are minimized then $f \sim 1$ and an incident particle will give a signal proportional to its energy.

Since the value 3·5 eV/electron-hole pair is generally about ten times lower than the energy/ion pair required in atomic ionization detectors, the statistical fluctuations in the number of charge carriers are reduced by a factor of about three, leading to resolution of around 1% in some cases.

This type of detector is useful in electron proton β- and α-radiation spectrometry. The width of the depletion layer used is determined by the type and energy of radiation of interest.

Lithium-Drifted Silicon Detectors

Lithium is an N-type donor in silicon, and the ions have a high mobility, since they move interstitially in the silicon lattice. Lithium ions may be "drifted" into P-type silicon at elevated temperatures by an electric field. The lithium is distributed to compensate exactly the acceptor concentration, forming an intrinsic (I), or impurity-carrier-free, region in the crystal. This I-region corresponds to the region depleted of charge carriers in the surface barrier diode by a reverse bias. A field may be set up in the Li-drifted device, by reverse biasing the N–I–P structure, but in this case the radiation sensitive volume is determined by the I-region and is almost independent of the magnitude of the bias.

Such detectors will stop a high energy particle such as a 25 MeV proton; they can also be used for β- and γ-spectroscopy. The relatively low atomic number of silicon ($Z = 14$) gives a small probability of photo-electric interaction and so gamma detection is mainly by Compton scattered electrons.

Neutron Spectrometers

This type of detector contains an element which produces charged particles when bombarded by neutrons. Two reactions are used. First, with ^{6}Li as the "converter" element, the reaction $^6\text{Li}(n, \alpha)\,^3\text{H}$ produces an alpha particle and a triton; secondly, with ^{3}He as converter, $^3\text{He}(n,p)\,^3\text{H}$ produces a proton and a triton. In each of these two types, the converter element is sandwiched by a pair of silicon surface barrier detectors. By coincidence counting techniques, one could, in principle, obtain a single peak for a monoenergetic beam, but in practice the background radiation is troublesome.

Chapter 5

SCINTILLATION SPECTROMETRY

D. GIBBONS

United Kingdom Atomic Energy Authority,
Wantage Research Laboratory, Wantage, Berks., England

Introduction

When a γ-ray interacts with a sodium iodide scintillator, it does so in two main ways. The most interesting reaction is the photoelectric process in which all the energy of the γ-ray is transferred to an electron and subsequently a flash of light is produced with an intensity essentially proportional to the incident energy. In the second process, the γ-ray is scattered, giving up only part of its energy according to the angle of scatter, producing a continuous distribution of energy transfer known as the Compton continuum. The resulting light flashes are converted to current pulses and amplified by a photomultiplier tube and the resulting distribution of signals is the γ-ray spectrum. A block diagram of typical equipment necessary to examine this pulse height distribution is shown in Fig. 13.

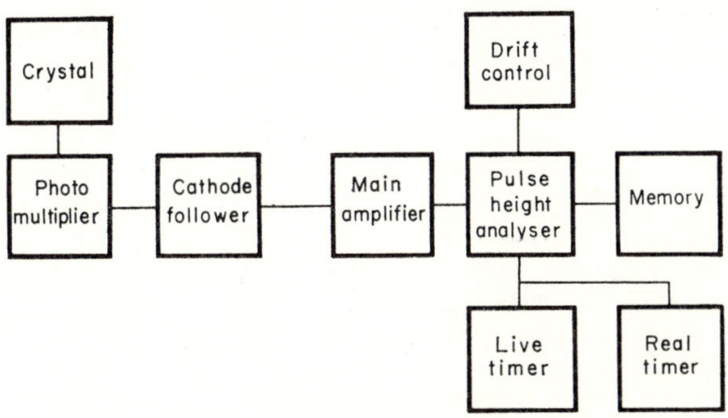

Fig. 13. Block diagram of γ-ray Spectrometer.

Cathode Follower

The cathode follower or pre-amplifier has a low output impedance; this enables a relatively long cable to be used between the photomultiplier and the subsequent electronic apparatus without the impedance of this cable being significant. The signal from the photomultiplier is a current pulse which, for a proportional response, must be converted to a voltage pulse by integration on a small capacitor. Because of the relation

$$V = \frac{Q}{C}$$

the output voltage from the capacitor is proportional to the charge input from the photomultiplier. It may also be seen that C should be small to give a large voltage pulse, but, on the other hand, it must be large enough so that its time constant is long enough to integrate the current pulse.

Main Amplifier

The function of this unit is to convert the signals to a size and shape suitable for the pulse height analyser. It should have an adjustable high gain (of the order of 10^3–10^4) which should be accurately calibrated; the gain characteristic should also be linear, so that a constant gain obtains over the complete range of pulse heights. The main amplifier must also have low noise characteristics, because the noise can interfere with the

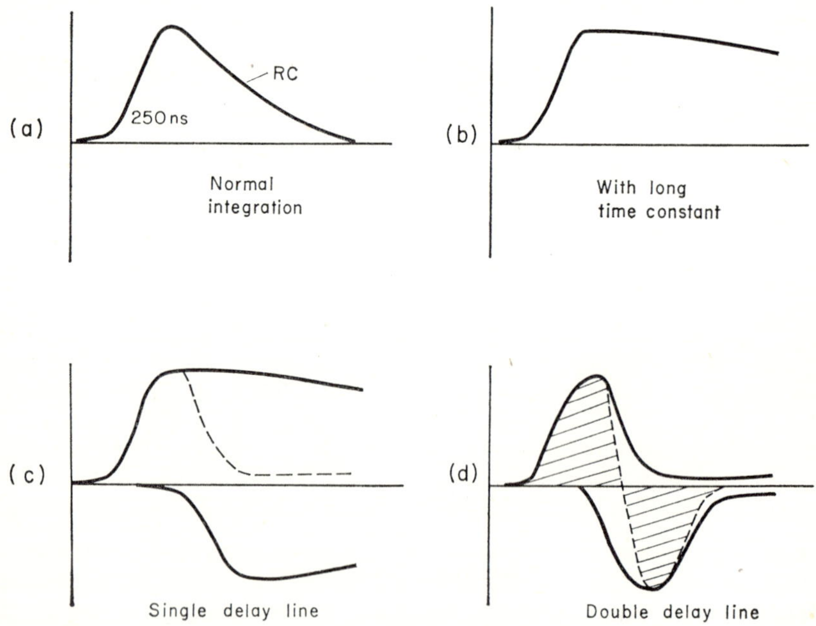

Fig. 14. Double delay line pulse shaping.

true pulses; since the noise is random, it increases the width of the photopeaks, even if the noise itself is cut out by the discriminator.

Occasionally very large input pulses (perhaps from cosmic rays) or very high count-rate bursts appear at the input of the main amplifier which will be "blocked" unless it has good overload characteristics. The former can be dealt with in two ways: the signal may be differentiated either by RC clipping, or by double-delay-line clipping. In the latter method the pulse is delayed for a very short time, inverted, and added to the original signal. The result is shown in Fig. 14c; a short pulse is obtained, but it has a longer base line or tail. Any other pulse that appears in this time will still be counted but will appear to have the wrong energy because the base line gives a false zero. The base line shift is therefore removed by delaying and further inverting the nett signal and adding it again. The result is a very short bipolar pulse as shown in Fig. 14d, which has good overload characteristics.

Pulse-Height Analysis

The distribution of pulse heights in the amplified signal can be examined in several ways. By using nothing more than a scaler in conjunction with a calibrated discriminator, the integral and differential

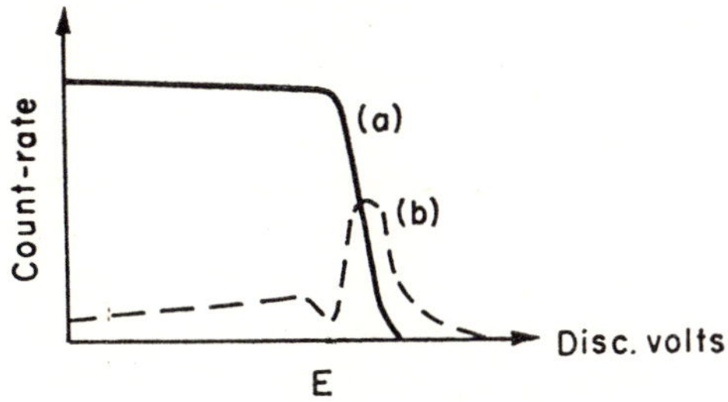

Fig. 15. Integral (a) and differential (b) spectra.

pulse-height distributions can be plotted (Fig. 15). If two discriminators are combined to form a single-channel analyser, this can be stepped through the spectrum in conjunction with a scaler, or swept through in conjunction with a ratemeter, to give the differential spectrum directly. The problem with all these methods, however, is one of time. With the emphasis in activation analysis continually moving towards shorter and shorter half-lives, the pulse-height distribution may well have changed before the full spectrum has been recorded. These limitations can be overcome by dividing the pulse-height range into a continuous series of channels and simultaneously recording the counts falling within the energy limits of each channel.

The Multichannel Analyser

Early multichannel analysers, therefore, consisted of a multitude of single-channel analysers. Although such systems had the potential advantage of short data storage times, they were extremely difficult to set up to avoid separation or overlap of adjacent channels and were susceptible to channel instabilities. In the more modern so-called computer-type of multichannel analyser the storage of data is preceded by analogue to digital conversion in which a number is generated in response to the size of each input pulse. This number, known as the channel address, then programmes a computer to store a count in the appropriate section of the memory of the analyser. This analogue to digital conversion is most commonly achieved by analogue to time conversion in which either the pulse is stretched into a flat-topped pulse

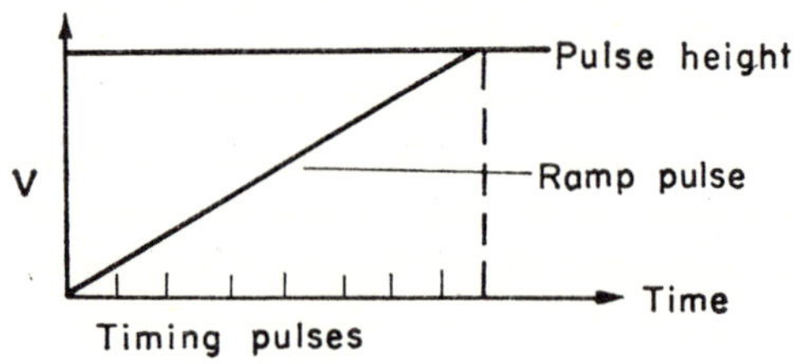

FIG. 16. Analogue–time conversion.

of the same amplitude and simultaneously an increasing sweep signal (Fig. 16) and a clock pulse generator are started, or a capacitor is charged to the peak pulse height and then discharged through a constant current device, turned on in conjunction with a clock pulse generator. In either case, the number of clock pulses occurring during the time required for the linear sweep to increase to the level of the stretched pulse, or for the voltage on the storage capacitor to discharge to zero, is counted to give the digital equivalent of the input pulse height.

Since it is inconvenient to have a separate scaler associated with each effective channel, alternative storage processes are used. These are of two types; periodic and non-periodic. Periodic storage systems are based on continuously rotating drums of magnetic material or loops of magnetic tape, or on regenerating ultrasonic electrical or acoustic delay line memories. In the latter, a train of pulses appropriately coded to represent the total content of all the channels is continuously regenerated and recirculated through the delay line. The ADC sweep voltage has the same period as the memory and is synchronized with it; consequently, although the count pulses may be random in time, they can only be stored at the appropriate point in the memory cycle. Since the input is normally gated to prevent the acceptance of a further pulse until the previous pulse has been sorted, the analyser will be "dead" from the time of receipt of the first pulse until the appropriate time in the memory cycle, and this will vary from zero up to the memory cycle time. It can be shown that the effective average dead time per pulse is, in fact, constant and equal to one-half the memory cycle time. Storage consists simply of modifying the pulse train to represent the new channel content, and readout of interpreting the pulse train.

In non-periodic storage systems, to be found particularly in the modern transistorized analyser, the memory is a static device consisting of a bank of magnetic cores which are magnetized appropriately to represent the information to be stored. In the simplest form, in which the channel count is stored in pure binary, there will be some 20 cores per channel to give a channel capacity of 10^6, so that a 256 channel analyser will involve some 5000 magnetic cores. The storage procedure is now somewhat different. When the channel address has been decided by the comparison circuit, the contents of that channel are transferred to a buffer which can also act as a scaler. The content of the buffer store is then increased by unity and transferred back into the memory. The dead time is now determined principally by the time required to determine the channel address, which will naturally vary from pulse to pulse from a few μsec to several hundred μsec depending on the clock pulse frequency and on the number of channels used for storage. The readout procedure involves transferring the memory content, channel by channel, into the buffer store and then decoding its content by some suitable serial or parallel readout device.

Because of the dead time of the multichannel analyser it is advantageous with periodic systems and virtually essential with non-periodic systems to use some means of measuring automatically either the total dead time or the live time associated with any counting period. If a stable oscillator is used to drive a time register and this oscillator is gated by the same waveform which gates the analyser input, oscillator pulses will only be counted when the analyser is "live", so that in a given total or real time, the register will indicate the live time or actual time available for analysis of pulses. Alternatively, if the live-time register is used to stop the analyser at a preset live time, then all counting periods will be for the same live time, regardless of the counting rate and resulting real time. However, in some situations, particularly when counting short-lived activities or mixtures of different half-lives, it is preferable to count for a fixed real time so that decay corrections are eliminated, and use the recorded live time to assess the count rate. The live-time oscillator must then run at such a rate that an adequate number of pulses are stored during the shortest real-time period to give the required accuracy in measuring the live time. In other circumstances it may be advantageous to record both live times and real times and control the analyser by the total number of counts stored in a channel or group of channels so that a standard statistical level is attained for each count.

The modern multichannel analyser is extremely versatile with regard to the additional features which have been incorporated. The memory can be split up into several sub-groups. In this way, several spectra from different detectors can be recorded simultaneously, or one spectra can be switched successively from sub-group to sub-group to give decay information without wasting time on readout between successive counts. The contents of one sub-group can be transferred to another and spectra can also be accumulated subtractively so that background subtraction can be accomplished. In some analysers arithmetic logic can be achieved in the transfer process so that some multiple of one sub-group or of some external information, can be added to or subtracted from another, so that unwanted peaks can be stripped out of the spectrum. Finally, most multichannel analysers can be switched so that all pulses are stored in channel 1, and then after a preset time all pulses are stored in channel 2, etc., thus giving, in effect "*n*" sequential scalers. This is particularly useful for decay studies of short-lived activities. A really versatile analyser is one in which a particular peak of interest can be selected by adjustment of upper and lower level discriminators, and then switched to "multiscaler" operation with only pulses between the preset pulse height limits being accepted.

Anti-Coincidence Shield

As will be seen, one problem associated with the evaluation of γ-ray spectra is the interference produced by the Compton continuum of high energy peaks in the measurement of lower energy peaks. The extent of this interference can be reduced by surrounding the normal detector with a large scintillator (e.g. NaI, CsI or a plastic or liquid scintillator) coupled to several photomultipliers to ensure efficient detection of all scintillations (Fig. 17). The combined pulses from the additional photomultipliers are fed into an anti-coincidence unit together with the signal from the central detector. If a Compton event occurs in the central crystal and the scattered γ-ray escapes, there is a good chance that it will be detected in the large outer scintillator. The anti-coincidence unit allows pulses to pass from the central detector for analysis only if they are *not* accompanied by a simultaneous pulse in the surrounding system.[1] In this way, photoelectric or total absorption pulses are analysed, but Compton pulses which are accompanied by an event in the outer detector, are not.

This system also has the advantage of reducing the cosmic ray background, since a background event in the central crystal will almost certainly be accompanied by an event in the outer detector. Unfortunately, γ-rays emitted in coincidence will also be reduced (including

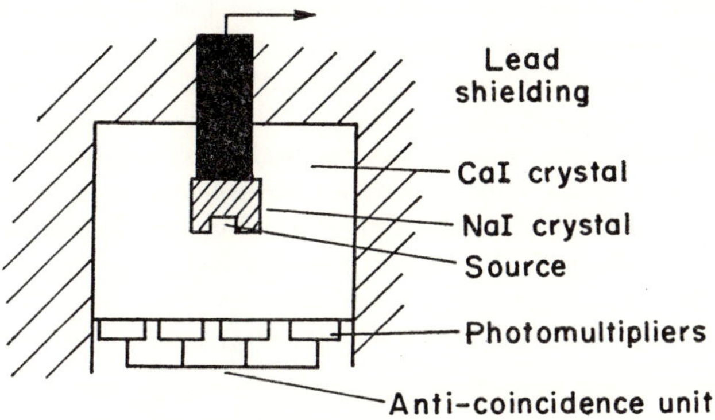

FIG. 17. Anti-coincidence detector.

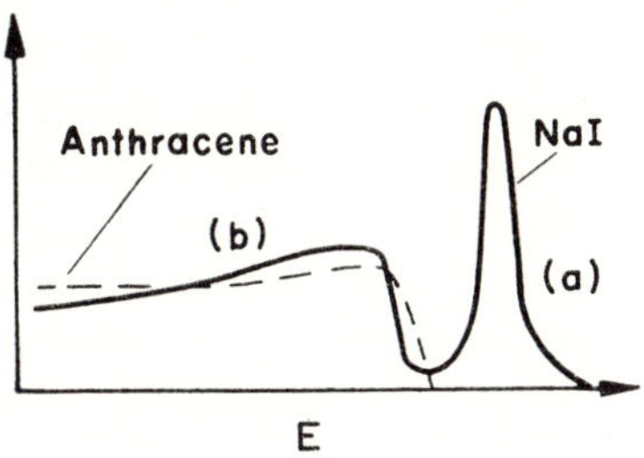

FIG. 18. Spectral response of sodium iodide and anthracene.

the 0·51 MeV γ-rays from positron emitters) and at high sample activities, random coincidences will also operate to reduce the effective sensitivity.

An alternative system, due to Peirson,[2] uses an anthracene or plastic scintillator detector in line with the source and sodium iodide detector. The effective gains of the two detectors are adjusted, and the position of the source between the two detectors also adjusted to give comparable Compton distributions (Fig. 18) in the two outputs. In the original version the two outputs were subtracted sequentially via a switching unit and differential ratemeter for single channel analysis. In conjunction with a modern multichannel analyser, however, the two outputs could be coded to give "dynamic" pulse by pulse subtraction within the analyser circuitry. This system has been modified by Hoste [3] who placed the two crystals at right angles and so eliminated the problems associated with back-scatter between the two detectors.

A further modification is due to Putman and Taylor [4] who used two completely independent matched sodium iodide detectors. The sample was counted via one detector and a mixture of appropriate quantities of all unwanted activities counted on the other. In this way, not only was the interfering Compton activity eliminated but all unwanted photopeaks as well. This system is, of course, a form of spectrum stripping and is better performed nowadays by the arithmetic logic of the modern analyser.

The problem in these subtraction methods lies in the matching of the two detectors, accurate thermostatting being necessary. In addition, in the Peirson/Hoste system, the detector efficiency is reduced as the sample must be closer to the second detector to give efficient compensation.

Multidimensional Analysis

In this system the sample is viewed by two detectors in coincidence and these detectors are contained in a third detector which serves as an anti-coincidence annulus for Compton and background suppression. The spectrometry system is a 4096 channel analyser used in a 64×64 multidimensional mode. The signals from the two detectors are fed to the two separate analogue to digital converters of the analyser for energy analysis and from there to the memory storage unit. When a single photon interacts with one of the detectors, the event is stored on the corresponding X or Y axis of the memory. When two photons are emitted simultaneously, each interacting with a separate detector, the event is stored in an energy–energy plane at a point uniquely characteristic of the two energies.

With a detection system of this type, the unique decay characteristics of many radionuclides are readily exploited to permit a far greater sensitivity for detection and direct measurement of many minor radioactive constituents. For example, the 8 radionuclides ^{46}Se, ^{60}Co, ^{64}Cu, ^{72}Ga, ^{106}Ru, ^{134}Sb, ^{140}Ba, ^{140}La can be measured in reactor cooling water, but could not be resolved by conventional spectrometry. In addition to the much greater selectivity provided by multidimensional analysis, a much lower background (100 to 1000 reduction) is obtained.

Energy Calibration

One of the most useful parameters which can be obtained in the calibration of the spectrometer is the relation between gamma ray energy and channel number called the channel coefficient and usually expressed in keV/channel. This coefficient only holds throughout the pulse-height range if the overall response is proportional ($y = mx$) or linear ($y = mx + c$) so that an estimate of the linearity is also required. Since the sodium iodide detector is itself non-linear below about 200 keV, it is often more convenient to measure spectrometer linearity with a precision pulse generator than with radioisotope standards.

It must be remembered that a plot of a multichannel spectrum is basically a histogram, so that the counts falling in channel 20 have an energy corresponding to channel 19·5 (or channel 20·5 depending on whether the first channel is called channel 0 or channel 1). If, however, the lower level and threshold controls are adjusted to set channel $-0{\cdot}5$ to correspond to zero energy, then a relative pulse height of 20 will fall in the centre of channel 20.

In the calibration of a spectrometer using radioisotopes, or in the estimation of unknown photopeak energies, it is important to be able to determine the position of the photopeak very precisely. Elementary methods such as measuring halfway along a horizontal line across the peak, or finding the intercept of the linearly extrapolated sides of the photopeak are regarded as less reliable than the capabilities of the system.

One method,[5] claimed to locate the peak to within 0·02 channels, defines the peak channel number PCN as the channel number which bisects the integral channel count, which in turn is defined as the integral count between the two channel positions corresponding to a count of one-half of the maximum in the peak.

Most other methods make use of the Gaussian distribution of the photopeak. By using the fact that the logarithm of a Gaussian distribution is a parabola, and locating the maximum of the parabola by plotting difference of logs of channel counts against channel number, and finding the zero intercept, the value of PCN can be found. Alternatively, one can plot the percentages of the channel counts compared to total photopeak counts against channel number on probability paper.[6] The channel number corresponding to 50% probability then gives PCN. Yet a further development plots the log of $y_{(x-1)}/y_{(x+1)}$ against x (i.e. channel number) on linear paper and measurement of x for $\log F(y) = 0$ gives PCN once more.[7] However, when statistics become important none of these plots is very good and some sort of fitting must be used. Unfortunately, one cannot use least squares fitting on these functions, since they are functions of the data points and it is only to the latter that the principle of least squares should be applied. However, since the Gaussian equation contains an exponential term, a linear fit cannot be made, and an iterative method must be used in which a guess is made of PCN and resolution, and corrections to these estimates are then computed by the method of least squares. This approach makes the use of a computer virtually essential.

Quality Control of Equipment

Unfortunately, the γ-ray spectrometer is subject to drift of its energy calibration even in the most carefully designed systems. This is due principally to count rate and temperature effects in the detector. Covell has compared different makes of photomultiplier tubes for their sensitivity to count rate induced drift and found that of the tubes examined at that time, the EMI 9578 was the best, but the effect was still present. The energy response of the detector is temperature dependent, and although thermostatted enclosures can improve the situation they do not represent a complete answer. However, by selecting a specific peak in the spectrum and interrogating the analyser to find in which channel the peak lies at any given instant, the calibration of the equipment can be checked.

Covell has devised a system of quality control[8] which is extremely useful. It is based on the principal that the ratio of the counts in each side of two reference photopeaks is measured and referred to a previously developed nomogram relating the two ratios to the setting of the high voltage and threshold controls. Thus, a few minutes spent each day in recording a spectra and evaluating the correction enables the spectrometer to be kept to a fixed calibration.

The gain of the equipment can be stabilized by means of a servo drift control system. The requirements of a drift control system have been discussed in the literature, but it is fairly obvious that a system which stabilizes the overall gain from γ-ray emitter to analyser memory is the best. Hence the reason for using one of the peaks in the spectrum as a reference. If the spectrum contains only unknown peaks, then a reference activity can be deliberately added, giving a high energy activity for control of gain and a low energy activity for the control of the lower threshold. Although dual stabilizers have been used on this principle, it now seems to be generally accepted that with the modern transistorized analyser, lower threshold drift is negligible, so only gain stabilization will be discussed here.

Two types of control circuit have been developed, one analogue and one digital. The former, in a variety of guises, compares the count rates in two groups of channels on either side of the reference peak in a dual ratemeter-type circuit and any out of balance voltage is used to drive a motor driven potentiometer to change the gain of the amplifier or the applied high voltage.[9] or may be applied directly to the high voltage stabilization circuit. In the digital system,[10] the total count in the two groups of channels is stored in an integral scaler, and the difference in a two-way differential scaler. The control limits are so arranged that if the peak drift is statistically significant, the differential scaler stops before the integral scaler, but if it is not, the integral scaler stops first. In the latter case the scalers reset themselves without applying any correction; while in the former case, the drift control operates a mechanism to change, for example, the photomultiplier voltage by one step. Thus correction is only applied when the count rate difference exceeds a preset statistical level, and the control system cannot drift itself since it is of a purely digital nature.

As has been stated already, if there is no convenient peak in the spectrum, a reference peak may be introduced deliberately, although this may then complicate the subsequent interpretation of the recorded spectrum. An alternative is to use a detector crystal containing a small quantity of a long-lived α-emitter; the 5–6 MeV α-radiation producing an equivalent 2·5–3 MeV photopeak without any accompanying Compton continuum. This arrangement is not entirely satisfactory, however, since the pulse height per MeV temperature coefficient of sodium iodide is different for α- and γ-radiation, and the difference depends on the method by which the crystal has been manufactured.

A more suitable procedure uses a separate plastic scintillator with a β-γ source embedded in it, the whole being situated close to the main detector. When β- and γ-rays are simultaneously emitted by this source, a fraction of the γ-rays pass into the original scintillation counter and give rise to a reference peak, while the β-scintillations operate an anti-coincidence unit which inhibits the "add-1" store process of the pulse-height analyser memory, but still allows interrogation of the channel address to provide the control information. The γ-ray pulses from the reference source are thus used as a drift reference peak but are not counted by the pulse-height analyser memory. The essential requirement of this system is that *all* β-radiation be detected to ensure that there is always an anti-coincidence pulse available for each unwanted γ-ray pulse. Thus, if only 99% efficiency can be achieved, then 1% of the reference spectrum incident on the main detector will remain in the recorded spectrum, but in most cases this will be negligible.

Measurement of Radioactivity

Comparative measurements are made by comparing the areas of corresponding photopeaks in two spectra. In a multichannel spectrum, the peak area can be determined by summing the individual channel counts over the photopeak region, if this is the highest energy peak in the spectrum (i.e. if the peak is sitting on a very low background). The situation is less simple in the presence of Compton continua from higher energy peaks. Here the area can be estimated by creating a base across the minima at either side of the peak, but this leads to subjective errors.

At a given resolution, the width at half-height of a peak is known. Consequently the area above the half-width is a function of the total area. This can be extended to the general case where the area above any fixed width (i.e. a set number of channels) can be related to the total area. Using the digital nature of the multichannel presentation this partial area can be calculated as:

$$\sum_{-n}^{+n} C_i - (n+\tfrac{1}{2})(C_{-n}+C_{+n})$$

where C_0 is the peak centre channel count.[11]

These approaches are limited to regions where the Compton continuum is reasonably flat.

A different approach involves graphical subtraction whereby the highest energy peak is identified, matched to a standard spectrum and then subtracted channel by channel throughout the energy range. This is then repeated until all the peaks have been identified and measured.

Commercial analyser accessory equipment is now available in which a variable amount of a standard spectrum may be subtracted from or added to a pulse-height distribution stored in the memory of the multichannel analyser using the internal logic processes already mentioned. This process has been called electronic spectrum unfolding or spectrum stripping. Use of a magnetic or paper tape read-in device is another means of spectrum stripping. An experimental spectrum stored in the analyser memory may be successively unfolded by repeated subtractions of standard spectra stored on tape. The only limitation is that each must have identical energy calibrations and identical zero-energy intercepts. Changes in amplifier gain in the experimental spectrum compared with that of the standard will cause errors in the resultant spectrum when the subtraction is accomplished.

Other techniques of instrumental data-reduction have been described. These methods consist in using the live-display feature of an analyser in a subtract mode. The unknown spectrum stored in the analyser memory is observed on the CRT, while a standard source of one of the constituents is allowed to subtract until its contribution to the source is minimized. If the time required for complete removal, and the disintegration rate of the standard are known, the amount of that nuclide in the unknown spectrum can be calculated. Individual components in a four-component mixture can be determined with precisions ranging from ± 1 to 30%. Precision is found to be poor for low-energy components in mixtures and is affected by the complexity of the spectral data.

Advantages and Problems

In conclusion it may be said that γ-ray spectrometry enables activation analysis to be used as a rapid non-destructive instrumental technique, with the possibility of determining several elements simultaneously. On the other hand, the full potentialities of the technique can only be exploited through the use of extremely complicated and often expensive electronic equipment.

References

1. Ellett, W. H. and Brownell, G. L. (1960). *Nucl. Instr. Methods,* **1**, 56.
2. Peirson, D. H. (1956). *Atomics,* **7**, 316.
3. Hoste, J. and Soete, D. (1961). Proc. 1961 Intern. Conf. on Modern Trends in Activation Analysis, p. 59.
4. Putman, J. L. and Taylor, W. H. (1958). *Trans. Soc. Glass Technol.* **42**, 84.
5. Dudley, R. A. and Scarpatetti, R. (1964). *Nucl. Instr. Methods,* **25**, 297.
6. Boekelheide, I. F. (1960). *Rev. Sci. Instr.* **31**, 1001.
7. Zimmermann, W. (1961). *Rev. Sci. Instr.* **32**, 1063.
8. Covell, D. F. (1963). *Nucl. Instr. Methods,* **22**, 101.
9. Fite, L. E., Gibbons, D. and Wainerdi, R. E. (1961). Proc. 1961 Intern. Conf. on Modern Trends in Activation Analysis, p. 102.
10. *Stabimat Spectrum Stabiliser,* High Volt Linear Ltd., Brussels 1964.
11. Covell, D. F. (1959). *Analyt. Chem.* **31**, 1785.

Chapter 6

ACTIVATION ANALYSIS—BASIC PRINCIPLES

R. E. WAINERDI

Texas A. & M. University, College of Engineering, College Station, Texas, U.S.A.

Activation analysis is based on the principle that when a material is irradiated by the nuclear particles produced in a nuclear reactor, particle accelerator or other suitable source, some of the atoms present in the material will interact with the bombarding particles and be converted into different isotopes of the same element or isotopes of different elements depending on the nature of the bombarding particles. In many cases, the isotopes produced are radioactive. If each different induced radioactivity can be distinguished or separated from all other radioactivities produced, then the amount of each radioactivity is a measure of the quantity of the parent isotope present in the material.

This basic principle is not new, being based on the work of Hevesy in 1936 and used to a limited extent by other workers prior to 1939. With the development of high flux reactors, during and following the Second World War, it became apparent that the technique of activation analysis offered considerable promise for the determination of trace elements owing to the high sensitivity which becomes available at high fluxes, and the technique developed quite rapidly. Later, Meinke and others showed that the technique was also of value in the analysis of macro-constituents using low-level neutron sources.

The estimation of minor elements in materials is often made difficult by contamination of the reagents. Impurity levels which can be accepted for most purposes become significant sources of error when the elements under investigation are present at levels of 1 p.p.m. or lower. For example, if one is searching for sub-microgram quantities of arsenic in a tissue specimen, any separation process involving sulphuric acid comes under suspicion because of the almost inevitable presence of arsenic in the acid.

This problem can be overcome by first activating the specimen to produce ^{76}As from ^{75}As; subsequent addition of ^{75}As as an impurity in the separation chemicals will have no effect on the activity, and in fact addition of ^{75}As will increase the accuracy of the experiment, for the following reason. Small quantities of As (or any other element) are easily adsorbed on to the walls of the glass beakers, etc.; thus the major proportion of activated ^{76}As may be lost in the process of separation. By addition of comparatively large amounts of ^{75}As we reduce the probability of the ^{76}As atoms being adsorbed, since ^{75}As and ^{76}As are indistinguishable for the process of adsorption.

Experiments carried out in this way can have sensitivities of the order of 10^{-9} to 10^{-12} g. A possible pitfall is the possibility that the ^{76}As atom may have come from ^{76}Se by (n,p) reaction or from ^{77}Se by (γ,p) instead of from ^{75}As as expected. These difficulties are unlikely when we irradiate the specimen with slow neutrons, since the $^{76}\text{Se}(n,p)^{76}\text{As}$ reaction has a high cross section for fast neutrons only.

As another example, consider the assay of ^{23}Na in ^{27}Al by $^{23}\text{Na}(n,\gamma)^{24}\text{Na}$. A competing reaction is $^{27}\text{Al}(n,\alpha)^{24}\text{Na}$ which gives the same amount of ^{24}Na as would the presence of sodium to the extent of 81 p.p.m. as an impurity in the aluminium sample. Accurate use of activation analysis thus involves elimination of interfering reactions by suitable choice of the exciting radiation.

For atomic numbers less than 10 or greater than 80 thermal neutrons are not very useful; for this reason H, C, N and O, which constitute 96% of living matter, are not activated by thermal neutrons. Use of fast neutrons may yield better results at low Z values.

Sources of thermal neutrons, e.g. research reactors, have thermal neutron fluxes of about $10^{12}\ n\ \text{cm}^{-2}\ \text{sec}^{-1}$ with about $10^{9}\ n\ \text{cm}^{-2}\ \text{sec}^{-1}$ of fast neutrons. These fluxes will give sensitivities in the microgram range, the most common reactions being (n,γ) for thermal neutrons and (n,p), (n,α) or $(n,2n)$ for fast neutrons. For elements of intermediate atomic number (between 10 and 80) the thermal neutron reaction is usually the most sensitive.

For example, aluminium may be estimated by four different reactions:

(*a*) Thermal neutrons: $^{27}\text{Al}(n,\gamma)^{28}\text{Al}$ $(t_{1/2} = 2{\cdot}3\ \text{min})$

(*b*) Fast neutrons: $^{27}\text{Al}(n,p)^{27}\text{Mg}$ $(t_{1/2} = 9{\cdot}45\ \text{min})$

(*c*) $^{27}\text{Al}(n,\alpha)^{24}\text{Na}$ $(t_{1/2} = 15\ \text{h})$

(*d*) $^{27}\text{Al}(n,\gamma)^{28}\text{Al}$ $(t_{1/2} = 2{\cdot}3\ \text{min})$

Typical sensitivities, based on a thermal neutron flux of $10^{12}\ n\ \text{cm}^{-2}\ \text{sec}^{-1}$, irradiation for four half-lives of the radioactive product or 1 h, whichever is shorter, and a final activity of 100 disintegrations per min (about the minimum that can be usefully detected), are as follows:

reaction (*a*)	0·02 μg
(*b*)	1 μg
(*c*)	17 μg
(*d*)	140 μg

Where possible the reaction selected by choice of excitation should be that which gives a fairly short half-life with the likelihood of a large count rate over a period of a few hours, which is enough for chemical and instrumental treatment.

Other methods of inducing activity include γ-rays (usually above 20 MeV) and protons. A useful proton reaction is $^{18}O(p,n)^{18}F$. The product has a half-life of 1·9 h. This reaction involves the introduction

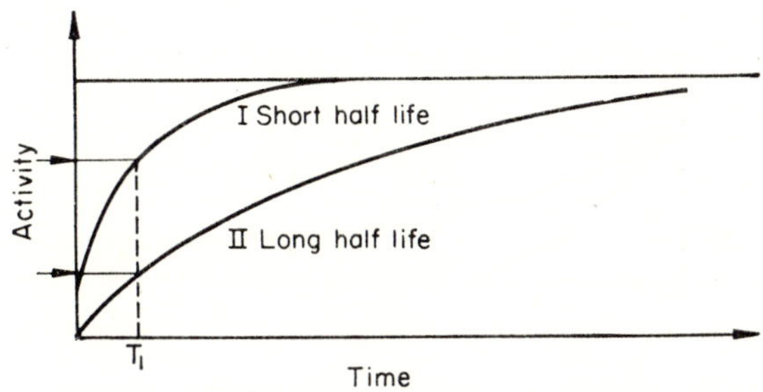

FIG. 19. Growth of activities for species with long and short half-lives.

of ^{18}O as tracer but gives good sensitivity in the analysis of biological material.

In practical activation analysis one is often presented with the problem of separating the activity of an element of interest from that of the contaminants or background. The first possible technique is to compare the spectrum of a mixture of active elements with that of a pure sample of the element of prime concern. If there is only one main contaminant, then comparison of the spectra will yield a certain amount of useful information, when reinforced by a check on half-lives of the main peaks.

If a sample contains two elements which have radioactive isotopes, then one can usually be reduced in importance if the half-lives are substantially different. In the activation curves shown in Fig. 19, it can be seen that by choosing a fairly short irradiation period, the activity of element I will be much greater than that of element II. In this way the estimation of I can be made with minimum interference from II. The actual irradiation time selected should be long enough to give an accurate count. If element II were of prime concern, then by choosing a long irradiation period, I would be saturated and II would be still approaching saturation. Then by postponing the count until I had decayed significantly, i.e. after four half-lives, the assay of element II could be carried out with relative ease (Fig. 20).

The problem of distinguishing the induced radioactivity due to one element from all other induced radioactivities can be approached in other ways. In what has been called "classical" activation analysis, the approach involves the complete radiochemical purification of the induced activities, separating the desired radioactivity completely from all other radioactivities. Such a purification is a tedious and expensive task and shows little prospect for automation or remote control.

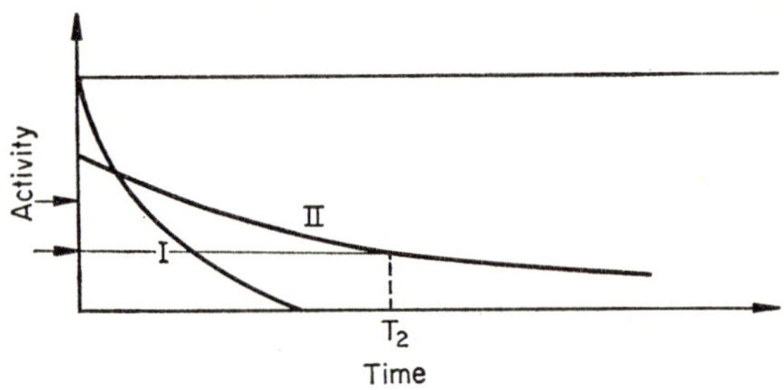

FIG. 20. Decay of short and long-lived species.

Another approach involves the use of the γ-ray spectrometer where radioactivities may be distinguished in terms of their γ-ray energy. Such a technique is excellent for qualitative identifications, but quantitative measurements on complex samples are difficult unless some sort of post-γ spectrometer data treatment is made to separate one γ-ray activity from another.

The digital computer is especially useful for this type of data analysis and computer programs have been written not only to resolve the γ-ray spectrum, but also to calculate the weight of the bombarded elements from their induced radioactivities. Such a procedure, for use in activation analysis, was described by Kuykendall, Wainerdi and Smith at Texas A. and M. in 1960, and other similar techniques are now available. In simple mixtures, manual analysis of the data recovered from an X–Y plotter can also produce qualitative and quantitative analyses.

Instrumental activation analysis may be employed to determine the qualitative and quantitative elemental composition of a material when the following conditions are satisfied:

(*a*) elements present in the material must have favourable neutron interaction cross-sections;
(*b*) each element to be analysed must produce one or more γ-ray emitting isotopes upon neutron activation;
(*c*) the γ-ray emitting isotopes formed as a result of the neutron bombardment must have half-lives which are suitably situated between convenient upper and lower limits;
(*d*) the elements to be analysed must be present in sufficient quantities; and,
(*e*) the nuclear characteristics of the non-radioactive element and its activation product(s) must be known.

If the above conditions are satisfactorily combined, a sample may be bombarded with 14 MeV neutrons for a predetermined length of time,

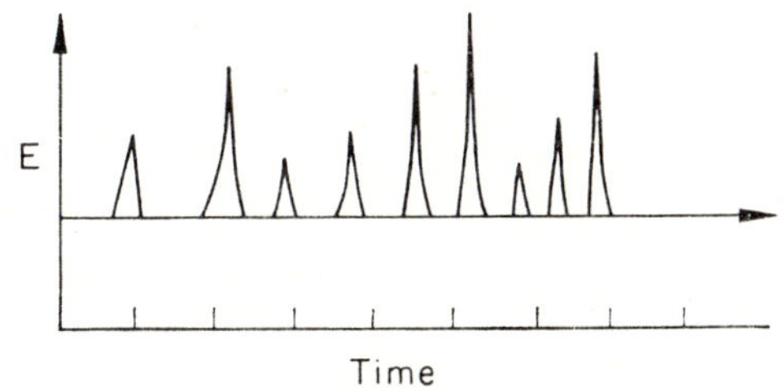

FIG. 21. Voltage pulses at the output of a photomultiplier tube.

counted with a scintillation spectrometer, and analysed for its chemical composition on the basis of its γ-ray spectra. To accomplish this analysis, a nuclear activation analysis system would require the following components:

(*a*) a source of activating particle (14 MeV neutrons);
(*b*) a detection system;
(*c*) timers and sample manipulators; and
(*d*) a data readout.

Figure 21 shows a series of voltage pulses at the output of a photomultiplier tube. These pulses are sorted by a multi-channel analyser and interpreted based on known nuclear decay schemes, and by comparison with standards of known composition. The crucial question underlying the reliability of such comparisons in the case of multi-component systems is applicability of the Law of Superposition.

The following article elaborates this matter

1. Wainerdi, R. E. and Du Beau, N. P. (1963). *Science*, **139**, 1027.

Chapter 7

NUCLEAR REACTORS AS SOURCES OF NEUTRONS

R. E. WAINERDI

Texas A. & M. University, College of Engineering, College Station, Texas, U.S.A.

Nuclear reactors are probably the most widely known neutron sources, and they are capable of producing by far the highest flux of neutrons (neutron flux is the number of neutrons passing per unit area per unit time). Neutrons are produced in the nuclear reactors through the process of nuclear fission, i.e. the breaking apart of heavy atoms, such as uranium or plutonium. In the fission of a uranium atom, two or three neutrons are produced. The energy released in each fission is approximately 200 MeV $= 200 \times 10^{6} \times 1{\cdot}6 \times 10^{-12} = 3{\cdot}2 \times 10^{-4}$ erg $= 3{\cdot}2 \times 10^{-11}$ joule. In other words, an energy release of 1 joule/sec corresponds to a fission rate of $3{\cdot}1 \times 10^{10}$ per sec. In order to determine how many fission neutrons are produced in a reactor operating at 5,000,000 W, it is therefore necessary to multiply $2 \times 3{\cdot}1 \times 10^{10} \times 5 \times 10^{6}$, which indicates that over 3×10^{17} neutrons are being produced per second. This is vastly more total neutrons than can be produced by any other presently available neutron source. However, many nuclear reactors are too large and expensive to be useful for many industrial applications; for this reason, other neutron sources producing fewer neutrons per second may be used in preference to reactors. It should also be noted that the neutrons produced in a reactor have a considerable range of energies. Fission neutrons can be found only slightly above thermal energy (0·025 eV) or with energies ranging up to more than 10 MeV. Since the probability of a neutron being captured in a particular material is dependent upon the energy of a neutron, it is quite difficult to use reactor produced neutrons for very precise analytical work (such as activation analysis) unless the neutrons are thermalized or otherwise sorted out as to energy. There are devices available which can sort reactor neutrons, but the number in any given energy level is generally less than the number which could have been obtained from a large, monoenergetic neutron source such as an accelerator.

Nuclear reactors are a unique form of prime mover in that they can be so constructed as to be independent of load (or heat removal rate) during the production of heat up to the point where the build-up of heat adversely affects the nuclear and structural properties of the reactor. For the purpose of comparison, a gas engine such as commonly used to power pumps can be loaded to a certain maximum. As the load is applied, the engine will slow down and labour, and finally, when the load exceeds the capacity of the engine, it will simply stop. Conversely, a "critical" nuclear reactor can be designed to operate at about any power, with the only limits being the ability of the operator to shield the machine and to remove the heat being produced before the reactor melts or otherwise loses something essential to the "chain reaction", such as the "moderator" material. In order to describe the chain reaction which occurs in nuclear reactors, the concept of "neutron economy" will be discussed.

In order for a nuclear reactor to remain in a "critical" condition and continue to chain react, the basic condition shown in Fig. 22 must exist,[1, 2] with as many thermal neutrons absorbed in the reactor as are being produced.

Consider 100 neutrons existing at the upper right-hand corner of Fig. 22, on the line just to the right of the box labelled "critical condition". Of these 100 neutrons, a certain number, assume 25, will be captured in the reactor's "poison" (or neutron absorbing) control rods, in the reactor's moderator (such as water or graphite) or in the reactor structural materials. This leaves 75 neutrons to be absorbed in the fissionable fuel. Some of these neutrons, about 60, will be captured in the fuel in such a way as to cause the fuel atom to split (or fission) and release, on the average, about two "new" neutrons, γ-rays, heat and two fragments of the former atom of fuel, which are radioactive debris called "fission products". The other 15 neutrons are captured in the fuel in a non-productive manner, and simply lead to the formation of a heavier atom of uranium. The 60 neutrons which did lead to thermal fission events fortunately give rise to about 120 fast (or high energy) neutrons. Most of the neutrons produced by the fission process have energies of about 1 MeV, but some of them are considerably more energetic, possessing up to 14 or 15 MeV. These fast neutrons can also induce a few fissions in uranium as they slow down, so about 15 more neutrons will be added from this source. The total number of neutrons produced by fission is, therefore, 135.

Neutrons diffuse very easily, and some will leak out of the chain reaction environment before being slowed down. About 5 neutrons will be lost at this point. During the slowing down process there is a non-productive and, therefore, entirely useless capture of neutrons in the

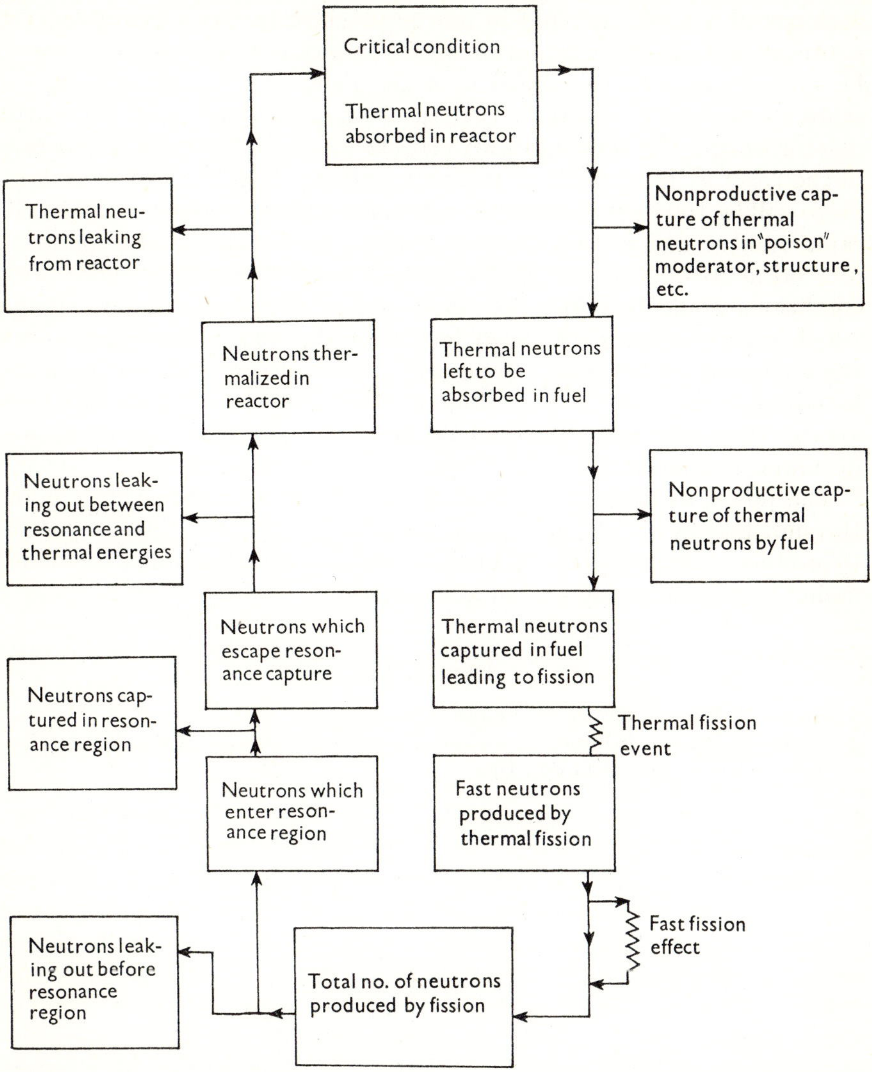

FIG. 22. Block-schematic representation of neutron balance in a reactor (from Glasstone and Edlund).

uranium fuel as the neutrons are slowed down to the point where their energy is equal to that of the "uranium capture resonances". At these particular energies (called resonances) there is a considerably enhanced probability that neutrons will be absorbed without the fission event taking place.

Since 130 neutrons survived to enter the resonance region, and perhaps 10 will be captured in non-productive resonance capture, 120 neutrons will escape resonance capture to become "thermalized" or to be slowed down to the speed that gas molecules would have at the temperature of the reactors. Perhaps 10 neutrons will leak out of the reactor during the slowing down process leaving 110 thermal neutrons in the reactor. If 10 of these thermal neutrons leak out of the system, there will be 100 neutrons left to be absorbed in the reactor, which is the same number which started on the cycle. This is considered a condition of good neutron economy, and a nuclear reactor operating under these conditions would be said to be "critical". Had the number of neutrons which survived the cycle been less than the 100 which started out, the nuclear reactor would be operating in a "sub-critical" (or reducing in power) condition. If the number of surviving neutrons had been greater than 100, the reactor would be in a super-critical (or increasing in power) condition.

The power of the reactor depends on how many neutrons are going through this cycle per unit time, for it takes 3×10^{10} fissions per sec to produce a watt of energy. A 1 million watt reactor therefore has 3×10^{16} fission events occurring within it each second.

References

1. Breazeal, Wm. M. Personal communication.
2. Glasstone, S. and Edlund, M. C. *The Elements of Nuclear Reactor Theory*. New York: D. Van Nostrand Co., Inc.

Chapter 8

USE OF ACCELERATORS AS NEUTRON SOURCES

V. P. GUINN

General Atomic Division, General Dynamics Corpn., P.O. Box 608, San Diego, California, U.S.A.

In activation analysis one has a choice between several incident activating particles. These fall into three categories:

(*a*) Neutrons, of thermal energies, or fast neutrons from the (d,t) reaction at 14 MeV. Neutrons of intermediate energies may also be used in special cases.

(*b*) Charged particles, usually of moderate energies, such as protons, deuterons, tritons, α-particles and ^{3}He nuclei.

(*c*) High-energy photons for photonuclear activation analysis through the (γ,n) and (γ,p) reactions, and occasionally through the (γ,γ') excitation reactions.

The choice of activating particles naturally depends on the particular samples to be analysed, and it is usually found that neutron-induced reactions cover a wider field than the other types.

Whereas neutron sources provide a fairly uniform flux over a large sample volume, the penetration of charged particles into matter is restricted to a surface layer typically less than 100 μ in thickness. This may be an advantage in particular cases, for example in analyses of thin surface films, micrometeorites and dust particles. Furthermore, charged particles of moderate energies cannot induce reactions in nuclei of high atomic number, because of the electrostatic Coulomb barrier which surrounds such nuclei. Charged particles of energies up to 6 MeV may be obtained from Van de Graaff accelerators and Cockcroft-Walton accelerators, while a tandem Van de Graaff generator raises the upper energy limit to 12 MeV. The use of higher energies extends the range of nuclei which can be activated to higher atomic numbers, but expense is usually an important limiting factor (the cost of a 12 MeV tandem generator is of the order of £200,000). Energies greater than 12 MeV may be obtained from a cyclotron. In this case it may even be necessary to slow the charged particles down to energies less than 25 MeV, because at these very high energies charged particles can cause very complex "spallation" reactions, usually of no interest to the activation analyst.

Reactions induced by ^{3}He nuclei are similar to those induced by α-particles, but generally with a threshold of about 15 MeV or less. An 8–10 MeV "table-top" cyclotron is very suitable for such things as carbon and oxygen determinations if ^{3}He is used. It is thus possible to detect perhaps as little as 10^{-11} g of oxygen, which is typically a factor of 1000 increase in concentration sensitivity over fast neutron analysis.

Van de Graaff generators, betatrons and linear accelerators, which are already available for electron acceleration, may easily be converted into photon sources for photoactivation analysis. The electrons strike a target of high atomic number, usually gold or tungsten, and produce

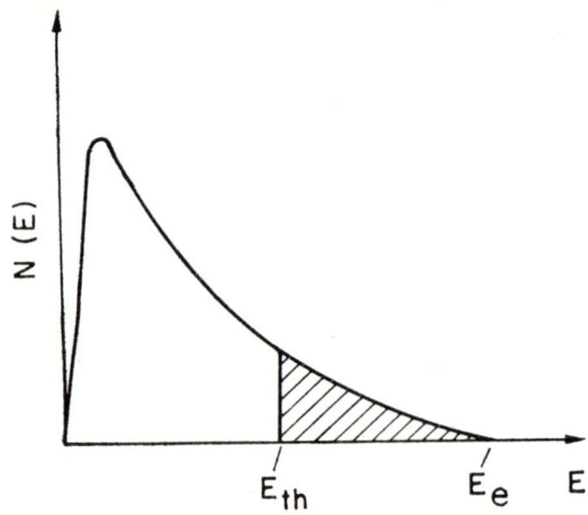

FIG. 23. Typical bremsstrahlung spectrum. (E is energy; N (E) is number of quanta of energy E.)

bremsstrahlung (deceleration radiation). Ten per cent of the energy of 3 MeV electrons is converted into photons in this way, and the efficiency rises with energy, so that 50% of the energy of 25 MeV electrons is converted into bremsstrahlung. The photons are also emitted preferentially in the forward direction, the anisotropy of the radiation increasing with electron energy. Figure 23 shows a spectrum of bremsstrahlung radiation from electrons of energy E_e. The low energy cut-off is due to absorption of the radiation in the target itself. It is unfortunate that most of the photons have too low an energy to be useful, but the "high energy tail" above the threshold energy, E_{th}, causes reactions of the types (γ,n), (γ,p) and (γ,γ'). The latter reaction is simply an excitation of the nucleus into an excited (isomeric) state, from which it subsequently decays with the emission of another γ-ray. It can be observed with several elements using the bremsstrahlung from a 3 MeV, 1 mA, Van de

Graaff generator, but the sensitivity is rather low. Higher beam currents are really necessary, and these can be provided by a 3 MeV "Dynamitron" accelerator which gives up to about a 10 mA beam, thus increasing the sensitivity ten-fold. Betatrons give electron beams of greater energy, typically between 10 MeV and 100 MeV, but usually with a very low beam current. A 45 MeV electron linear accelerator can provide a maximum beam power (13 kW) at 0·5 mA and 25 MeV—higher energies are obtainable at the expense of beam current.

An electron accelerator may also be easily converted into a neutron source for neutron activation. As shown in Fig. 24, a beryllium block is placed directly beyond the gold or tungsten converter. Bremsstrahlung photons interact with the beryllium nuclei and produce low-energy neutrons. A 3 MeV Van de Graaff electron accelerator can make available

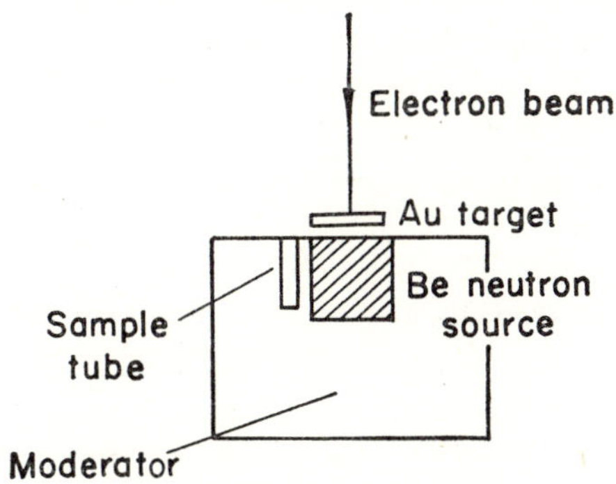

FIG. 24. Neutron source arrangement.

fluxes of 10^8 thermal neutrons cm^{-2} sec^{-1} in this way. The fluxes obtainable from tungsten targets with a 25 MeV linear accelerator may be as high as 10^{10}–10^{11} n cm^{-2} sec^{-1}, which are nearly as high as those found in nuclear reactors. These fluxes are high because neutrons are produced in the tungsten itself, as well as in the beryllium, through a (γ, n) reaction. The simultaneous production of neutrons and γ-rays by one target means that photoactivation experiments are not easy to carry out with tungsten targets; a neutron flux is always present, and the sample will usually have a larger cross-section for neutrons, causing interferences. A reverse interference effect occurs in reactors, where there is always a γ-ray flux. The γ-interference in a reactor is, however, negligible because the γ-rays generally have only low energies.

It is thus possible to use an electron accelerator to produce thermal neutron fluxes comparable with the fluxes in a small research reactor. Positive ion accelerators can also produce these high fluxes; but for fluxes in the range above 10^{12} n cm^{-2} sec^{-1} a reactor is considerably more economical.

Many types of small neutron generators using positive ion acceleration are now commercially available. Cockcroft-Walton generators using a 1 mA beam of 150 keV deuterons produce fluxes of 10^9 n cm^{-2} sec^{-1} at the sample location, through the (d,t) reaction. These neutrons have an average energy of about 14 MeV; they may be moderated to thermal velocities with paraffin wax or other hydrogenous material. The tritium–titanium targets decline in output rather rapidly and a constant high flux cannot normally be maintained for a very long period of time. A 2 MeV Van de Graaff generator, giving deuteron beam currents of 150 μA, can also be used. By using a beryllium target, which has an indefinite life, thermal-neutron fluxes of about 10^9 n cm^{-2} sec^{-1} can be maintained.

The large number of elements which can be detected using thermal neutrons is shown in Tables I and II, which include the limit of sensitivity at various neutron fluxes. A flux of 10^5 n cm^{-2} sec^{-1} may be obtained from isotopic sources, e.g. a plutonium–beryllium mixture; while fluxes of 10^{12} n cm^{-2} sec^{-1}, and higher, are obtained in a research-type nuclear reactor.

TABLE I

THERMAL NEUTRON LIMITS OF DETECTION FOR 65 ELEMENTS

Neutron flux $= 2 \times 10^{12}$ n cm^{-2} sec^{-1}
Irradiation time $= t_i \leqslant 1$ h

(For γ-detection only, interference-free basis)

Sensitivity limit μg	*Elements*									
5×10^{-6}	Dy									
1×10^{-5}										
5×10^{-5}	Lu	Mn								
1×10^{-4}	Ag	Ho	In							
5×10^{-4}	Au	Eu	Rh							
1×10^{-3}	Cu	Er	Ir	Re	V					
5×10^{-3}	As	Br	Ga	La	Na	Sm	Sr	U	W	
1×10^{-2}	Al	Hg	I	Sb	Yb					
5×10^{-2}	Gd	Ge	Pd	Pr	Ru	Sc	Te	Th	Ti	
1×10^{-1}	Ba	Ce	Cl	Co	Mo	Nd	Pt	Tb	Tm	Zn
5×10^{-1}	Cd	Cs	K	Mg	Ni	Sn	Ta			
1	Cr	F	Hf	Nb	Zr					
5	Ca	Rb	Se							
10										
50										
100	Fe	S								
500	Si									

TABLE II

NUMBER OF ELEMENTS DETECTABLE INSTRUMENTALLY AT VARIOUS THERMAL NEUTRON FLUXES

Irradiation time = $t_i \leqslant 1$ h

(For γ-detection only, interference-free basis)

Limit of detection	*Number of elements detectable at flux of*					
	2×10^{12}	2×10^{9}	2×10^{8}	2×10^{7}	2×10^{5}	n cm^{-2} sec^{-1}
$\leqslant 1\ \mu$g	59	14	6	1	0	
$\leqslant 10\ \mu$g	62	28	14	6	0	
$\leqslant 100\ \mu$g	64	47	28	14	1	
$\leqslant 1$ mg	65	59	47	28	6	

Use of the tables. Example:

Table II shows that 14 elements are detectable in quantities less than or equal to 10 μg with a flux of thermal neutrons of 2×10^8 neutrons per square centimetre per second. Table I has been calibrated for a flux of 2×10^{12} n cm^{-2} sec^{-1}, which is greater by a factor of 10^4. One therefore moves a factor of 10^{-4} up the sensitivity limit column in Table I, i.e. from 10 μg to 10^{-3} μg. The elements in this row and in the rows above are the 14 detectable elements referred to in Table II, namely Dy, Lu, Mn, Ag . . . Ir, Re, V.

Elements which form only pure β^--emitting products have been excluded, as have been a few elements (e.g. O, N, P) which are detected instrumentally only by fast-neutron activation.

Chapter 9

VAN DE GRAAFF GENERATORS AS NEUTRON SOURCES

W. BYGRAVE

High Voltage Servicing Co. Limited,
Eastcote, Middlesex, England

In 1930 R. J. Van de Graaff invented the electrostatic high-voltage machine which now bears his name. It can be used to accelerate protons and other charged particles to moderate energies to initiate nuclear reactions and, in particular, neutron-producing reactions. The energy of the charged particle beam can be controlled to a few kilovolts, and the electrical current of the beam may also be held constant to within a few per cent. The other main advantage of this type of accelerator is that it can be made quite compact, by surrounding the high voltage terminal with a pressure vessel containing a gas insulant.

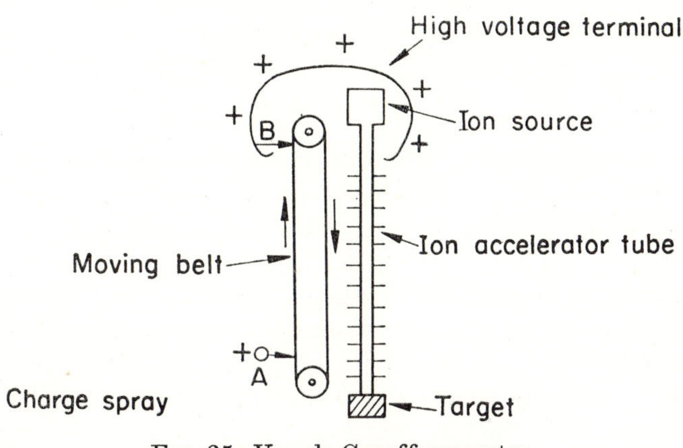

FIG. 25. Van de Graaff generator.

The principle of operation of the Van de Graaff generator may be understood by reference to Fig. 25. A voltage is applied from an external source to the point electrode A, and charge is deposited by a corona discharge on to a moving insulated belt. This carries the charge up to a further point electrode, B, which collects the charge and transfers it to the high voltage terminal. If this terminal has a capacitance C, then it will acquire a voltage

$$V = \frac{Q}{C}$$

where Q is the amount of charge transferred by the moving belt. In this way it is possible to obtain voltages of several millions of volts. Positive ions (protons, deuterons, etc.) are accelerated from an ion source at this voltage along an evacuated accelerator tube to an earthed target. Clearly the choice of positive ion and the choice of target depend on the particular nuclear reaction which is desired. For the production of neutrons, reactions are commonly of the (d,n) type, in which a deuteron beam is accelerated at a target nucleus, which then emits a neutron, increasing its atomic number Z and its mass number A by one unit. Neutrons can also be produced using protons, α-particles, or γ-rays, as incident particles: the (p,n) reaction only becomes useful at energies in excess of about 2 MeV.

The most important of all nuclear reactions in low voltage (< 400 keV) neutron generators is the tritium–deuterium reaction. These are both heavy isotopes of hydrogen, consisting of a proton with two neutrons, and one neutron, respectively. The reaction in which they participate may be written

$$^{3}_{1}\mathrm{H} + ^{2}_{1}\mathrm{H} \longrightarrow ^{4}_{2}\mathrm{He} + ^{1}_{0}\mathrm{n}$$

or,

$$^{3}\mathrm{T}(d,n)\,^{4}\mathrm{He}$$

The tritium is embedded in titanium metal to form a target of about 2·5 mg cm^{-2} thickness. These targets have a rather short lifetime and it is important to recognize that the neutron flux from such a target will decrease with usage. The number of neutrons per second from a titanium–tritium target may be reduced by about a factor of two after 1 mAh of operation. This is partly offset by cooling the target with a water jacket. The heating effect may also be reduced by scanning the beam spot over the target by a deflector electrode, by rotating the target, or by defocusing the ion beam. The short lifetime of the target is also due to deuterons settling in the titanium and diluting it and displacing the tritium, and of course by the tritium being used in the reaction itself. It has been found possible to accelerate tritons into the target as well as deuterons, and this can increase the target life very considerably. Unfortunately, tritium is a β-emitter with a half life of approximately 12 years, and the ion source and the accelerator may become contaminated with this unwanted activity.

TABLE I

PARAMETERS OF H.V.E.C. VAN DE GRAAFF GENERATORS, TOGETHER WITH NEUTRON OUTPUTS

Model	*Energy (max)*	*Current (max)* μA	*Tank length*	*Neutron yield neutrons/sec* $\times 10^{10}$			*Thermal neutron flux at 2 cm from target*	
				$^3T(d,n)$	$^9Be(d,n)$	$^7Li(p,n)$	$^3T(d,n)$	$^9Be(d,n)$
AN 400	400 keV	150	4′ 1″	3·0	0·06	—	$1{\cdot}6\times 10^7$	$0{\cdot}6\times 10^7$
AN 2000	2 MeV	150	7′ 7″	4·0	5·0	—	$1{\cdot}6\times 10^7$	$5{\cdot}0\times 10^8$
KN 2000	2 MeV	750	9′ 11″	13·3	37·5	—	$5{\cdot}3\times 10^7$	$3{\cdot}7\times 10^9$
KN 3000	3 MeV	400	11′ 8″	11·3	60	60	$4{\cdot}3\times 10^7$	$6{\cdot}0\times 10^9$
KN 4000	4 MeV	400	15′ 6″	11·3	120	120	$4{\cdot}3\times 10^7$	$1{\cdot}2\times 10^{10}$

(1) Thick targets used for Be and Li. 2·5 mg cm^{-2} T–Ti Target.
(2) Thermalization in a 24-in paraffin cube.
(3) All $^3T(d,n)$ yields calculated at 0·8 MeV except for the AN 400 which is at 0·4 MeV.

At energies of 2 MeV and above the ^{9}Be$(d,n)^{10}$B reaction is the most commonly used source of accelerator neutrons. The beryllium target is an elemental metal and has an almost indefinite lifetime which is an outstanding advantage compared with titanium–tritium targets.

The Van de Graaff neutron generator normally produces in the (d,t) reaction neutrons of approximately 14 MeV. These may be thermalized by surrounding the target with a paraffin-wax moderator. This, unfortunately, increases the target–sample distance and the flux is reduced in accordance with the inverse square law. It should be remembered that the target is essentially a point source of neutrons which are ejected more or less equally in all directions.

An advantage of this type of generator is that it may be pulsed to give short sharp bursts of neutrons. This is extremely useful when dealing with sample elements of short half-life.

The Van de Graaff accelerator may also be used for activation analysis by direct bombardment with positive ions.[1] The abundances of the elements Li, Be, B, C, O, F, Na, Mg, Al and P can all be measured after activation with 2 MeV protons in a Van de Graaff accelerator. The 7 MeV γ-rays from the reaction ^{19}F$(p,\alpha)^{16}$O*, ^{16}O* $\rightarrow$ ^{16}O $+ \gamma$ have a high enough energy to be detected free of Compton background radiation, and this method is the most promising one for fluorine determination. The analysis of carbon in steels is easily carried out with the reaction ^{12}C$(d,n)^{13}$N, ^{13}N $\rightarrow \beta^{+} +$ ^{13}C. Positive ions have a very short range in matter and only a thin surface section of the sample is activated. The activation may be restricted to specific areas of the sample by focusing the ion beam to a spot of perhaps 1 mm diameter. Positive ion analysis has the further advantages that no sample transfer is necessary, and so the prompt activity of a short-lived isotope may be measured; and also that the Coulomb barrier surrounding nuclei restricts reactions to elements with a low atomic number. Thus to a certain extent one can select the range of elements that are being measured and activate those only.

Table I shows some parameters for typical Van de Graaff generators with various target reactions. The neutron energy is 14 MeV for the ^{3}T$(d,n)^{4}$He reaction and between 1 and 6 MeV for the ^{9}Be$(d,n)^{10}$B reaction; as the ^{7}Li$(p,n)^{7}$Be reaction is endothermic, it does not occur until the incident protons have an energy of at least 1·9 MeV. The neutron energy then increases as the proton energy is stepped up.

Reference

1. Sippel R.F. and Glover E.D. (1960). *Nuclear Instruments and Methods,* **9**, p. 37.

Chapter 10

FAST NEUTRON ACTIVATION ANALYSIS

D. GIBBONS

United Kingdom Atomic Energy Authority,
Wantage Research Laboratory, Wantage, Berks., England

Although some of the early work on activation analysis was accomplished through the use of particle accelerators, it was the development of the high-flux nuclear reactor and the simultaneous need for materials of high purity which led to the technique of neutron activation analysis being developed as a highly sensitive and specific method of trace analysis. More than 60 elements can be determined to better than 10^{-7} g/g and 30 to better than 10^{-10} g/g at fluxes greater than 10^{12}. Since fluxes are currently available approaching 10^{15}, the sensitivity of the method is obviously outstanding.

Reactor activation analysis is equally applicable in the non-trace element range, but has been little used for this purpose, principally because of the disadvantages in time and cost compared to other methods of analysis, especially for the analyst without ready access to a reactor. Attempts to use radioisotope neutron sources (e.g. Ra/Be, Po/Be, Sb/Be, etc.) for laboratory activation have proved virtually unsuccessful because of the very low neutron fluxes produced ($\sim 10^8$–10^9 output). Fortunately, in recent years several relatively inexpensive particle accelerators have become available and can be used as useful sources of neutrons for a range of analyses where rapid operation at a relatively low cost, rather than high sensitivity, is the most important consideration.

Accelerator Neutron Sources

These can be divided into two main types, depending on whether electrons or positive ions are accelerated. The electron Van de Graaff, or linear accelerator, can be made to produce neutrons via photo disintegration reactions. The electron beam from the accelerator is made to produce high energy X-rays, by interaction with a high atomic number target such as gold, and the X-rays in turn produce neutrons by causing photo disintegration of a secondary target of Be (1·66 MeV threshold) or of Uranium (9 MeV threshold). These reactions produce neutrons of 1–2 MeV mean energy and therefore tend to be used as means of producing thermal neutrons as they can be moderated fairly easily. With a beryllium target, about 9 in. cube, little additional moderation is required, and the maximum thermal flux is obtained about 2 cm from the primary target.

Positive ion accelerators are probably the more useful systems for neutron production. Once again, beryllium can be used, but now as a primary target for the $^{9}Be(d,n)^{10}B$ reaction. This requires a bombarding energy greater than 1 MeV (e.g. 2–3 MeV Van de Graaff) and produces neutrons of 2–5 MeV. Thus, this reaction is again favoured for thermal neutron production. On the other hand, by using a titanium tritide target, and the $^{3}H(d,n)^{4}He$ reaction, 14 MeV neutrons are produced efficiently at beam energies in the 0·1–1·0 MeV range. This can therefore be achieved with the low-energy positive-ion Van de Graaff, or with one of several 150 kV accelerators.

14 MeV Neutron Generators

Since the operation of all such machines is basically similar, a typical design will be described.[1] The generator consists essentially of an ion source, an acceleration tube and a target, Fig. 26. Positive ions of

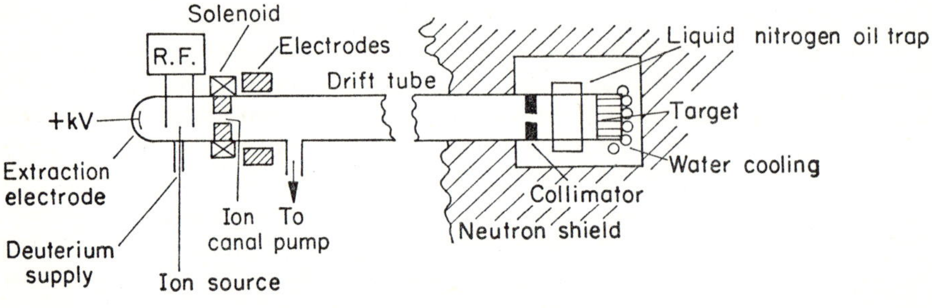

Fig. 26. Construction of a neutron generator.

deuterium are produced in the ion-source, often by radio-frequency excitation, and are then forced out of the ion-source by the extraction potential to be focussed and accelerated in the beam tube.

It is not possible to produce a pure beam of d^{+} atomic ions, and there will always be a large number of molecular d_2^{+} ions present. Owing to their larger mass, these have less effective energy and result in less efficient neutron production. The deuterium efficiency is also very low, i.e. the ratio of the number of deuterons in the ion beam to the number of deuterium atoms introduced into the source is small. The ratio of the number of neutrons produced to the number of deuteron ions in the beam is again small, and in these terms the D-T neutron generator is very inefficient.

The neutron flux for a thin target varies with deuteron energy as shown in Fig. 27, reaching a maximum at about 110 keV. The beam energy should clearly exceed this, to give a high intensity neutron source; in fact it should be arranged so that the beam energy has been attenuated to this level (by target absorption) at about the centre of the target. Acceleration potentials of 120–150 kV are used depending on the thickness of the titanium–tritium target.

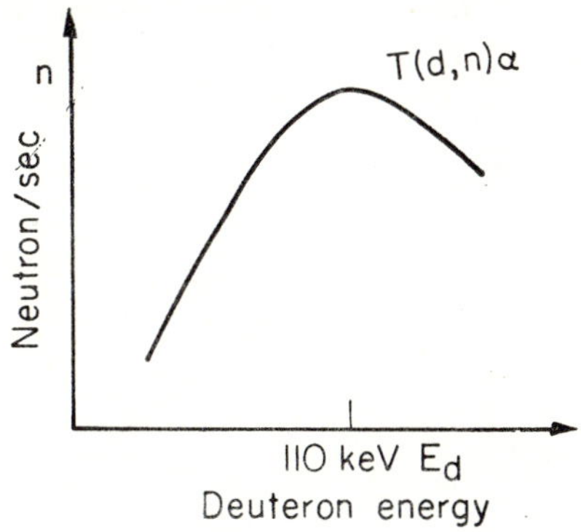

FIG. 27. Neutron flux as a function of deuteron energy.

In the majority of 150 kV neutron generators, the beam tube is continuously pumped, and deuterium is supplied continuously, at a controlled rate via a heated palladium leak, to the ion source. Older models use conventional oil diffusion pumps, with liquid nitrogen traps, etc., but more modern machines use getter-ion pumps, and more recently the latter has been made in line with, and part of, the beam tube so that more efficient pumping results.

However, regardless of the type of accelerator, the target is the limiting feature of the system. Typical target assemblies give an output of 10^{10}–10^{11} neutrons per sec, but this decays by a factor of 2 after 1–2 h, Fig. 28. Thicker targets are claimed to give longer lives by providing a tritium reservoir or, by using higher acceleration potentials to give increased penetration, higher outputs should be obtained. It has been claimed that rare earth–tritium targets, by being more thermally stable, should give increased target life. However, this does not seem to have been borne out in practice, probably because the fall-off is due to sputtering of target material rather than simple loss of tritium. Other attempts to combat the falling output have involved the use of rotating target assemblies. These have given constant outputs up to 20 h, but generally involve less favourable sample-target geometries.

More recent developments involve the use of dual-beam tubes to enable targets to be recharged.[2] When the neutron output has fallen to a suitable level, the deuterium supply is replaced by a tritium one and the target bombarded with tritium to replenish it, this operation being achieved at a lower accelerating potential.

A rather different type of generator, giving an output of 10^{10} n/s, is now commercially available.[3] This is completely sealed, and operates with a mixed beam of deuterium and tritium, thus giving constant replenishment and constant output for periods of up to 100 h.

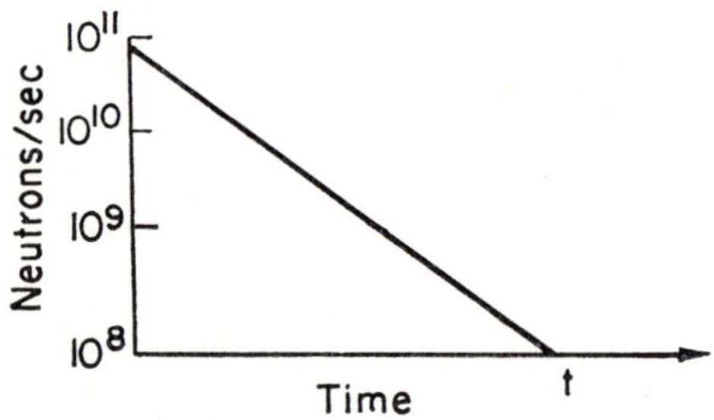

FIG. 28. Decay of target yield with time.

Although this has a lower output than the initial output of a pumped generator, the output of the latter soon falls.

The output of the sealed 10^{10} tube is compared with a typical 6×10^{10} pumped tube in Fig. 29. It is possible to keep the neutron production rate of the latter constant by starting at a flux below the maximum and adjusting the accelerator controls to give a constant neutron flux, as shown by line (a), until the output can no longer be maintained. Alternatively, starting at maximum output the target could be changed when the averaged output has fallen to 10^{10}. In either case, the target life is 8–9 hours, and the running costs of the two machines are then comparable. Consequently even though the whole sealed tube needs to be replaced after 100 h it is still more convenient to use, provided the output is adequate. It is expected that sealed tubes giving 10^{11} n/s will soon be available and these would then tend to be the preferred method of 14 MeV neutron production.

All generators capable of producing useful fluxes must be provided with an effective radiation shield. A concrete cube, 10 ft overall and

4 ft thick, leaving a 2-ft cube at its centre for moderation or experimental equipment is considered adequate for outputs up to 10^{11}. The main body of the generator is housed outside the cube and the target is situated at the end of a long "drift" tube.[4] This can introduce focussing problems, but is probably the most convenient method of operation. On the other hand, the 10^{10} sealed tube is about 7 in. diameter by 4 ft long in its container and can be very readily sited down a hole in the ground.

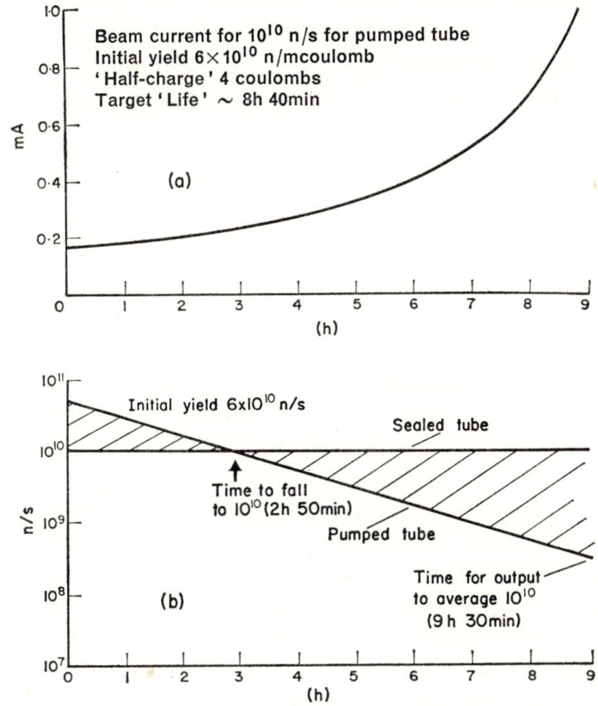

FIG. 29. Comparison of pumped and sealed generator outputs.

Activation Analysis Systems

In order to make use of short-lived radioactivities to take advantage of the accessibility of these machines, sample transfer systems are normally employed. These are usually pneumatic, of either the pressure or flow type and, typically, have transit times of a few seconds. Very high speed systems such as those suggested for use with pulsed reactors might have some application. Although calibrated dual transfer systems can be used to allow the conventional simultaneous irradiation of sample and standard, this is not always convenient and it is more usual to irradiate the two separately using some sort of instrumental neutron monitor for intercalibration. The beam current itself varies with time, with the accelerator voltage, and with the state of the ion source, and cannot be relied upon for calculation of the neutron production rate, but any direct or indirect neutron detector can be used for monitoring, and although special detectors have been used for specific purposes, the most generally used detectors are fission chambers of BF_3 counters, coupled to suitable ratemeters or scalers. Neutron monitoring is, however, complicated by the fact that, as the target is effectively a point neutron source, the neutron density gradient in its vicinity is very steep, and the relation between flux at the sample and flux at the monitor may vary.

The chief problem in neutron monitoring is in making allowance for variations in flux during irradiation. This can be done mathematically, or also electronically by using an integrating ratemeter with a time constant set to correspond to the decay constant of the isotope being measured. If the neutron output is turned off at the end of irradiation, and the ratemeter read at the beginning of the sample count, this method also takes care of differing decay times as well.

The most convenient method of determining the induced radioactivity will be via γ-ray spectrometry since, with short-lived radioisotopes, there is little time for radiochemistry and, for industrial operation, a purely instrumental procedure is more desirable. The modern transistorized multichannel analyser is capable of handling high intensity, rapidly decaying counting rates, and has many additional features to simplify the problem of interpreting the spectra, although computer unscrambling has many additional advantages. In some cases, the predominant induced activity is restricted to positron emission. Spectrometry can be used to measure the associated 0·51 MeV γ-ray but, since many 14 MeV produced activities also emit positrons, it is necessary to resort to coincidence counting and decay curve analysis with computerized interpretation as appropriate.

A complete activation analysis system of generator, shield, transfer system and counting system can probably be purchased and assembled for about £15,000 if a little home construction and ingenuity is employed.

Activation Analysis

THE DETERMINATION OF OXYGEN

The 14 MeV neutron activation method for the determination of oxygen has received considerable attention in recent years as it appears to be a most promising instrumental method for rapid analyses. The nuclear reaction involved is

$$^{16}O(n,p)\,^{16}N \xrightarrow[7\cdot4\ \text{sec}]{\beta^-} {}^{16}O$$

and is followed by the emission of γ-rays of about 6 and 7 MeV. Apart from fluorine, which interferes directly by ^{16}N production through the (n,α) reaction, boron is the only common element which can give rise to γ-rays of comparable energy. However, in certain matrices, where high intensity, low energy γ-rays are produced, it may still be difficult to measure low intensities of high energy γ-rays. In determining oxygen in hydrocarbons, the only other radioactivity produced is ^{13}N formed in secondary reactions (e.g. $^{13}C(p,n)\,^{13}N$) with knock-on protons. Thus, simple discriminated γ-ray counting procedures can be used and, in favourable cases, even total-count methods may be adequate.

THE DETERMINATION OF ALUMINIUM IN A COPPER ALLOY

The aluminium content of a copper–aluminium–iron alloy can be determined from the ^{27}Mg activity induced by the reaction $^{27}Al(n,p)\,^{27}Mg$.

The neutron flux and irradiation time for each sample are chosen to yield about the same activity due to aluminium, copper and iron in the standards as expected in the alloy, based on its nominal composition. The spectrum of the irradiated aluminium sample will show the 0·84 and 1·02 MeV γ-rays of 9·5 min ^{27}Mg, and these γ-rays are also in the spectrum of the alloy. There are no γ-rays in this energy region in a pure copper spectrum but γ-rays of energy 0·84 MeV from ^{56}Mn, produced by the reaction $^{56}Fe(n,p)\,^{56}Mn$, can be expected due to the iron content of the alloy. The aluminium content of the alloy can be calculated using the areas of the 0·84 and 1·02 MeV photopeaks independently. A correction to the area of the 0·84 MeV photopeak can be made based on the nominal iron content of the alloy, by reference to the iron spectrum.

Since the alloy has a density significantly different from that of the aluminium standard, it is necessary to consider the effects of γ-ray absorption during counting. This can be calculated assuming exponential absorption and neglecting γ-ray emission from the sides of the sample.

THE DETERMINATION OF PHOSPHORUS AND BARIUM IN LUBRICATING OIL

Several additives are used in modern lubricating oils to improve or modify the properties of the oils, and the analysis of these additives is therefore of industrial importance. The neutron generator has a possible application as a routine analytical tool in this field, since some of the elemental additives undergo convenient interactions with 14 MeV neutrons.

Examination of reaction cross-section data for phosphorus and barium indicate that the most sensitive reactions for their determination are:

$$^{31}\mathrm{P}(n, \alpha)\,^{28}\mathrm{Al} \xrightarrow[2\cdot3\ \mathrm{m}]{\beta^-} {}^{28}\mathrm{Si}: \quad \sigma = 146\ \mathrm{mb}$$

$$^{138}\mathrm{Ba}(n, 2n)^{137\,\mathrm{m}}\mathrm{Ba} \xrightarrow[2\cdot6\ \mathrm{m}]{\gamma} {}^{136}\mathrm{Ba}; \ \sigma = 1250\ \mathrm{mb}$$

In this way, phosphorus and barium can be determined, rapidly and non-destructively, in lubricating oil in the range 0·01–1% and 0·05–2% using 14 MeV neutron activation and γ-ray spectrometry. Practical limits of measurement are probably of the order 0·004% and 0·001%.

References

1. Prud'homme, J. T. (1962). *Texas Nuclear Neutron Generators*. Texas: Texas Nuclear Corp., Austin.
2. Hollister, H. (1964). *Nucleonics*, 68.
3. Wood, J. D. L. H., Lomer, P. D. and Bounden, J. E. (1963). Proc. 3rd Congèrs International de Biologie de Saclay, p. 119.
4. Olive, G., Cameron, J. F. and Clayton, C. G. (1962). A.E.R.E. R-3920.

Chapter 11

NEUTRON ACTIVATION ANALYSIS WITH REACTOR PULSES

V. P. GUINN

General Atomic Division, General Dynamics Corpn., P.O. Box 608, San Diego, California, U.S.A.

In the normal steady-state nuclear reactor, control is effected by the insertion or removal of control rods of neutron-absorbing materials which reduce the reactivity of the core. In a reactor which can be pulsed, such as the General Atomic TRIGA, the core does have control rods inserted in it for steady operation, but one of these is also used for initiating a high-intensity reactor pulse. In this type of reactor, the fuel elements themselves consist of a metal hydride, which serves as a moderator as well as a fuel (U-Zr-H). As the power level rises after rapid removal of the main control rod, the supercritical core becomes heated and hence the minimum (thermal) energy attainable by the neutrons rises. Since fission reactions in the fuel have lower cross-sections for higher energy neutrons, the reactivity will drop with increasing temperature. Thus the reactor flux will rise rapidly until it reaches the temperature at which an average of less than one neutron per fission causes another fission and the reactor automatically goes sub-critical and shuts itself off. The control rod is then reinserted leisurely, to prepare for the next pulse.

In the TRIGA reactor, some 70 cylindrical fuel elements, consisting of 92% zirconium and 8% uranium (of which 20% is ^{235}U), partially hydrided, are assembled as a cylindrical core as shown in Fig. 30. The individual fuel elements are jacketted in aluminium or stainless steel.

Whereas the first pulsed system of this type attained a power level of about 300 MW, as against 0·01 MW, in the steady state, essentially the same system now provides 1000 MW pulses and 0·25 MW steady-state operation. The pulsed-action power level (1000 MW peak) corresponds to a peak total neutron flux of $3{\cdot}1 \times 10^{16}$ neutrons per cm^2 per sec. The shape of the pulse is as shown in Fig. 31, the full width at half-height being about 20 msec.

When a sample of a given material has been irradiated in a reactor,

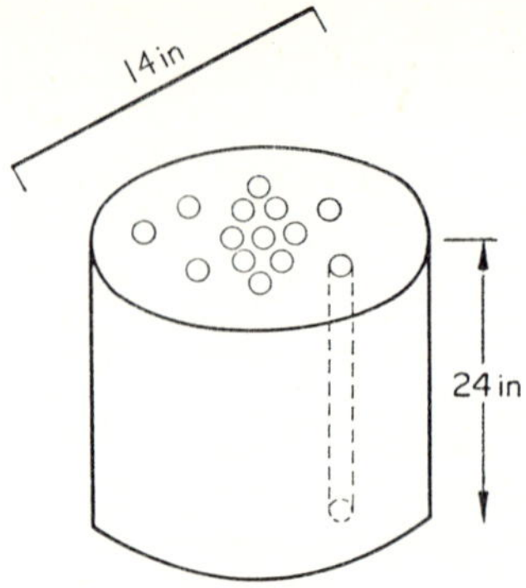

FIG. 30. Representation of a TRIGA core, with 70 fuel elements arranged in concentric circles.

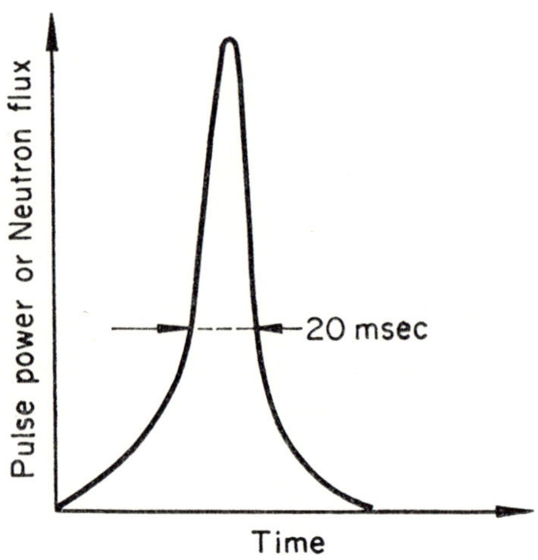

FIG. 31. Shape of reactor pulse with time.

the induced activity, of a specified type, at the time of removal is given by,

$$A_0 = N\phi\sigma S$$

where S is the fraction of saturation attained, and is written as

$$S = 1-e^{-\lambda t_i} = 1-\exp[-(0{\cdot}693 t_i/t_{0\cdot5})]$$

Here N is the number of target atoms of the appropriate type, ϕ the neutron flux, σ the cross-section for the required reaction, t_i the duration of irradiation, and $t_{0\cdot5}$ the half-life of the product nuclide.

For exponent values of the order of $< 0{\cdot}05$, this can be approximated by,

$$A_0 = N\phi\sigma\frac{0{\cdot}693t_i}{t_{0\cdot5}}.$$

In steady-state reactors, the value of S is approximately 1 for materials having half lives short compared with the duration of the irradiation. If one writes the activation of an element producing a short half-life product as

$$A_0^1 = N\phi^1\sigma\frac{0{\cdot}693t_i}{t_{0\cdot5}},$$

for a pulse, and the corresponding saturation activation in steady-state operation as

$$A_0 = N\phi\sigma$$

then the ratio of activation level reached in a pulse to that at normal steady-state saturation is:

$$\frac{A_0^1}{A_0} = \frac{0{\cdot}693t_i^1\phi^1}{t_{0\cdot5}\phi} \approx \frac{50}{t_{0\cdot5}}$$

(for a 1000 MW pulse). For nuclides with long half-lives, this pulse technique has no advantage, but with short half-lives one can enhance the detection sensitivity by a factor of $50/t_{0\cdot5}$ (the half-life being expressed in seconds).

In an experimental plot of the ratio A_0^1/A_0 against $t_{0\cdot5}$, the lecturer found this to be nearer $70/t_{0\cdot5}$ than $50/t_{0\cdot5}$, using a least-squares-fit method.

Thus, for products of half-life less than 70 sec, the activity is increased by using a pulsed output: 700-fold for $t_{0\cdot5} = 0{\cdot}1$ sec, 70-fold for $t_{0\cdot5} = 1$ sec, 7-fold for $t_{0\cdot5} = 10$ sec. It might be thought that the interference from long half-lived products would be greatly reduced, but since we are in the linear region of the activation curve this does not follow.

The question of modification of the energy distribution of the neutrons does arise, but since the thermalization time is around 10 μsec, and the pulse duration is in the millisecond range, no change is detectable. With the TRIGA Mark F reactor at San Diego, normally operated at 1·5 MW steady power, pulses as high as 6500 MW are now produced—giving peak neutron fluxes $> 10^{17}$ n/cm^2-sec.

The list of elements having short-lived activation products includes,

O, F, Na, Mg, S, Se, As, Sc, Y, Tb, W, Pt and Au; of these, oxygen and selenium have the following reactions, which are fairly typical:

	$t_{0\cdot5}$
$^{18}O(n,\gamma)\,^{19}O$	29·4 sec
$^{16}O(n,p)\,^{16}N$	7·35 sec
$^{76}Se(n,\gamma)\,^{77m}Se$	17·5 sec
$^{77}Se(n,n')\,^{77m}Se$	17·5 sec
$^{78}Se(n,2n)\,^{77m}Se$	17·5 sec

In general, radioactive isotopes may be formed in one or more of the reactions (n,γ), (n,p), (n,α), (n,n'), $(n,2n)$. A few of the short lived nuclides are β-emitters in the 10 MeV range, much higher energies than are usually obtained from β-radiation. Using a Cerenkov counter consisting of water or plastic in a container coupled to a photomultiplier tube, high β-count rates have been recorded with negligible interference from γ-radiation.

The method of transporting the sample to and from the reactor core is adapted to suit the particular requirements of short half-life analysis. The rabbit can be fired pneumatically into the core, where it triggers a pulse and then travels round a loop, receiving radiation for the duration of the pulse: it then leaves the reactor and describes a closed loop around the counting crystal, gradually slowing to a halt. The time of irradiation is so short that the sample does not suffer any damage by heating.

The TRIGA reactor used by the lecturer is being used at present in such studies as the determination of previously unknown half-lives and the reassessment of those which have been measured by less favourable techniques by earlier workers.

Chapter 12

RADIOCHEMICAL METHODS IN ACTIVATION ANALYSIS

S. J. THOMSON

Chemistry Department, The University, Glasgow, W.2., Scotland

In the ideal instrumental method of activation analysis, chemical techniques are unnecessary. However, in many cases, interfering or contaminating nuclides must be removed by chemical methods before the counting process can begin. When chemical methods are used, it becomes necessary to consider the radiochemical yield, that is the percentage recovery of the wanted nuclide. This can be measured by the use of a carrier technique during the recovery process. The appropriate equation is:

Per cent yield of added carrier
= per cent yield of product radioactive atoms.

We must also consider the time element when chemical separations are required: all chemical steps should be as rapid as possible if the wanted nuclide is short lived.

An important concept which arises when chemical methods are used in activation analysis, is that of the number of atoms involved. Normal chemical procedures involve milligram (mg) quantities of material, i.e. 10^{19}–10^{20} atoms. In activation analysis, the number of radioactive atoms to be separated might be equivalent to 10^{-10} g. If the species has atomic weight 100, then $10^{-10}\text{ g} \equiv 6 \times 10^{11}$ atoms. In a glass container with 10^{15} adsorption sites/cm^2 it is then possible to lose all the radioactive atoms on these sites. The simplest precaution to avoid such losses is to wash the container with special reagents. This is not particularly satisfactory, and a much better method exists. This involves the addition of carriers.

To the target material containing species a^*, b^*, c^*, . . . and w^*, where w^* is the species to be isolated, a^*, b^*, etc. are contaminants, we add A, B, C, . . . and W mg of these species. If the carrier contains 10 mg, or about 6×10^{19} atoms of W, then the loss on 10^{15} sites becomes 0·002%. Should precipitation reactions be required in chemical separations, the presence of carriers means that we may, if we wish, exceed solubility products and make precipitates in the normal fashion.

The addition of carriers thus aids chemical separations in more ways than one. Chemical behaviour, when carrier atoms are absent, becomes unpredictable, since w^* can be adsorbed or co-precipitated on unlikely precipitates. Carrier-free chemistry can therefore be considered unsatisfactory from the separation point of view. The obvious remedy is to add carriers to *all* our chemical systems. This raises the problem of the stage at which they should be added and in what quantity? The answer to the first is that carrier is added at the earliest possible stage in separation processes. For liquid samples, the solution of carriers should be added before any chemical techniques are applied to it at all; for solids, addition is made immediately before digestion. As to how much is added, 3–100 mg are the approximate limits: 10 mg is the normal amount. Too much is undesirable if the sample is to be counted by β-decay, for self-absorption in the final sources prepared for counting may become significant. An exceptional case arises where ion exchange processes are involved, for a limit to the amount of carrier is necessary if we are not to saturate the resin.

An interesting case arises where there are no suitable carriers available, e.g. for neptunium-239, which has no stable compounds. It is then necessary to resort to the periodic table and choose an element with similar chemistry. Lanthanum would appear to be suitable in this case. A much more elegant method is to use the α-emitter neptunium-237, since the radiochemical yield can be estimated from α-emission.

In an activated sample, we shall assume that a^*, b^*, c^*, . . . and w^* radioactive atoms have been produced and that A, B, C, . . . and W mg of carrier have been added. Our ultimate aim is to produce a particular element $w^* + W$, in a suitable form, free from all other contaminants in good yield in as short a time as possible. This can be done in two ways; (i) precipitation of $w^* + W$ in some form, leaving $a^* + A$, $b^* + B$, $c^* + C$, . . . in solution: A, B, C, . . . are then known as "hold-back" carriers. Repetition of the precipitation is required to remove any unwanted adsorbed species; (ii) precipitation of $a^* + A$, $b^* + B$, $c^* + C$, . . . leaving $w^* + W$ in solution. In both methods, the procedure should preferably contain a specific chemical step for the final isolation of $w^* + W$ only (e.g. solvent extraction, volatilization).

So far, nothing has been said of the chemical form of the irradiated atoms. In a nuclear transformation, there is the possibility of chemical, as well as nuclear, change. Consider the neutron-irradiation of chlorine-35 in ethyl chloride:

$$^{35}\mathrm{Cl} + n \begin{array}{l} \nearrow\ ^{36}\mathrm{Cl} + \gamma \\ \rightarrow\ ^{35}\mathrm{S} + p \\ \searrow\ ^{32}\mathrm{P} + \alpha \end{array}$$

It is necessary that we should consider the chemical state of the new atoms. Is the chlorine-36 substituted in the parent ethyl chloride, or does it appear as a free atom? What is the valency state of the sulphur and phosphorus atoms? In the neutron interaction with ethyl iodide a thermal neutron will interact with the iodine of the organic molecule, with the consequent production of a new nuclide,

$$^{127}\mathrm{I}(n,\gamma)\,^{128}\mathrm{I}$$

I*, the active iodine produced, can be extracted into an aqueous medium. This is quite unlike the behaviour of the iodine in normal ethyl iodide, which is not extracted into water. Some chemical (as well as nuclear) reaction would therefore appear to have taken place. The mathematical equation for the process of recoil of the product ^{128}I atoms is,

$$E_{\mathrm{recoil}} = \frac{536E_\gamma^2}{A}$$

where E_{recoil} = energy of recoil of iodine atom (in eV),
E_γ = energy of the γ-photon (in MeV),
A = atomic weight of the element.

Substituting appropriate values $E_\gamma = 4{\cdot}8$ MeV, $A = 127$ (for iodine), then

$$E_{\mathrm{recoil}} = 96 \text{ eV}$$

Since the strength of the typical chemical bond is a few electron volts, we can expect that such a bond will be readily broken by the ^{128}I recoil, with consequent release of an atom of iodine. As well as the displacement of iodine atoms, secondary reactions can occur, e.g. substitution,

$$CH_3CH_2I(n,\gamma) \xrightarrow{\text{recoil}} I^* \quad \text{(displacement)}$$

$$I^* + CH_3CH_2I \xrightarrow{\text{substitution}} CH_3CH_2I^* + I \quad \text{(substitution)}$$

We can therefore conclude that the products of irradiation of ethyl iodide are varied, but there are two distinct possibilities: (i) release of I*; and (ii) retention of I* in the organic molecule. We must therefore expect that our wanted element w^* in activation analysis may turn up in unusual forms.

The most general conclusion that we can draw is that there is no simple fate for an atom on irradiation, e.g.

Organic molecules,

CH_3Cl, CH_3Br, CH_3I $\longrightarrow$ release of I, Br, Cl respectively, followed by substitution

Inorganic molecules,

$$ClO_3^-, BrO_3^-, IO_3^- \longrightarrow \text{release of Cl, Br, I respectively}$$

$$SeO_3^=, SeO_4^= \longrightarrow \text{Se and Se}^{IV} \text{ respectively}$$

$$Au^{III}, As^{III}, Se^{VI} \longrightarrow \text{Au, As, Se respectively}$$

Valency reduction is usual in inorganic materials, but this is not a rigorous guide.

It is now necessary to consider what happens to these unusual chemical forms after irradiation when the specimens are dissolved, for example, in water. If we take permanganate ion as an example,

$$MnO_4^- \xrightarrow{\text{irradiation}} MnO_3^+ + MnO_2^{3+} + MnO^{5+} + Mn^{7+}$$

then hydration,

$$MnO_3^+ + 2OH^- \longrightarrow MnO_4^- + H_2O \quad \text{(alkaline solution)}$$

or reduction by water,

$$4MnO_3^+ + 2H_2O \longrightarrow 4MnO_2 + 3O_2 + 4H^+$$

can occur according to pH. From these reactions it is obvious that the fate of the irradiated species can depend upon the conditions of dissolution.

We can now draw two general conclusions regarding radiochemical methods in activation analysis. These are: (i) we must add carrier; and (ii) we must use an oxidation or reduction process to get all the w^* atoms and the W carrier atoms into the same chemical form.

The most useful radiochemical separations are tabulated below.

Method	*Minimum Time (minutes)*	*Selectivity*	*Concentration Dependence*	*Applicability*
Volatilization	1	good–excellent	No	non-metals
Solvent Extraction	1	moderate–good	No	most elements
Chromatography	30	good–excellent	No	most elements
Precipitation	2	poor–good	Yes	most elements
Electrodeposition	30	moderate–good	No	some
Mass Spectrometry	10	excellent	No	all

All of these methods, with the exception of mass spectrometry, depend on a distribution of the elements between two phases, e.g.

Solvent Extraction—partition between two liquids, one water, one organic solvent,

Ion Exchange—partition between solid phase and aqueous phase,

Paper Chromatography—partition between stationary paper and moving liquid.

Because of this distribution dependence it is appropriate to consider some definitions.

(i) The distribution of a solute A between two phases, 1 and 2, at equilibrium, is given by the distribution coefficient D_A

$$D_A = [A_1]/[A_2]$$

where [ ] represents total concentrations irrespective of chemical state. D_A need not be a constant.

(ii) Separation factor,

$$\alpha = D_A/D_B$$

for the separation of two species A and B. If $\alpha = 1$, then there is no separation. For efficient separation α deviates significantly from unity. This can usually be accomplished using adjustment of pH, chelation, etc.

(iii) The distribution law is given by

$$p_A = [A_1]/[A_2]$$

p_A is a constant at low concentrations for species with the same molecular weight in each phase. The difference between the distribution coefficient and the distribution law should be appreciated. As an example of this law we can consider the distribution of gallium between ether and 6 M hydrochloric acid; this has a constant value of $p_A \simeq 18$ in the range 10^{-3} M–10^{-12} M. The important concept, from the practical viewpoint, is the distribution coefficient D_A defined as

$$D_A = \frac{\text{Total concentration in organic phase}}{\text{Total concentration in aqueous phase}}$$

This ratio, which is what the analyst wants, varies with pH, etc. We must explain this behaviour, and to do so we shall consider the distribution of osmium tetroxide between carbon tetrachloride and water. In water, we have,

$$OsO_4 + H_2O \rightleftharpoons H_2OsO_5$$

$$H_2OsO_5 \rightleftharpoons H^+ + HOsO_5^-$$

$$HOsO_5^- \rightleftharpoons H^+ + OsO_5$$

Thus the distribution ratio

$$= \frac{[OsO_4]+4[(OsO_4)_4] \quad \text{(organic)}}{[OsO_4]+[H_2OsO_5]+[HOsO_5]+\ldots \quad \text{(aqueous)}}$$

The second term in the numerator is from polymerization in the organic solvent.

The second and remaining terms in the denominator are pH dependent, thus the distribution ratio can vary with pH.

Having considered radiochemical separations in the general terms, we shall now examine some of them in greater detail. Let us consider the various factors affecting solvent extraction. Of fundamental importance is some guide to solubilities. Covalent compounds, possessing low melting points and low lattice energies, are soluble in covalent organic solvents, e.g. iodine in chloroform, benzene in cyclohexane. Ionic substances have high lattice energies, and so are insoluble in organic media, but are soluble in water, liquid ammonia, and liquid sulphur dioxide, where there is the possibility of solvation of ions, and the dielectric constant of the medium renders ion separation easier. Fortunately, ions can be rendered soluble in organic solvents, by the formation of uncharged species from charged ions. This process can be accomplished in two ways.

The first of these processes is the neutralization of charge by a purely electrostatic type of bonding; e.g. rhenium can be made soluble in organic media by use of tetraphenyl arsonium perrhenate

$$\{[Ph_4As]^+[ReO_4]^-\}$$

As further examples, complex uranium species

$$\{(UO_2)(NO_3)_2(\text{ether})_4\}$$

$$[(UO_2)(\text{isobutanol})_6]^{2+}[(NO_3)^-]_2$$

can be extracted into ethers or isobutanol. Oxonium extraction also falls into this class—ferric iron can be extracted from hydrochloric acid into ether as a complex etherate,

$$\{[Et_2O:H]^+[FeCl_4(Et_2O)_2]^-\}$$

The second process for changing charged into uncharged species is that of chelation. In general we can write

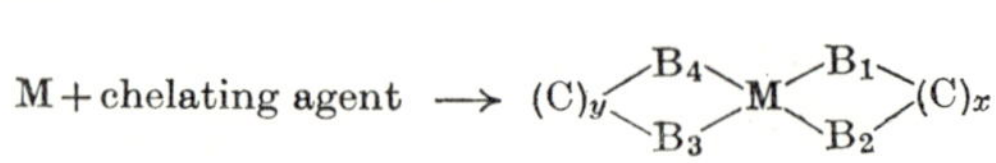

where M = metal, C = carbon atoms, B_1, B_2, B_3, B_4 = O, S, or N, x and y are generally 5 or 6. Typical examples of these chelates are:

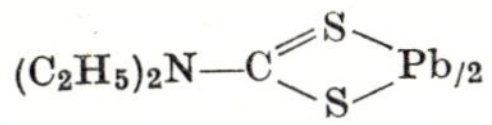

Lead diethyldiothiocarbamate

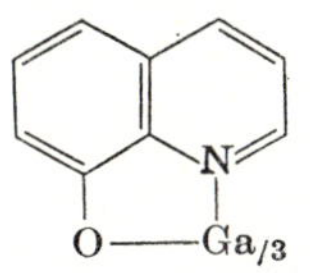

Gallium *p*-quinolate

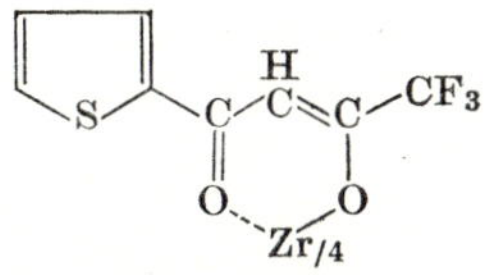

Zirconium thenoyltrifluoroacetonate

Two further techniques must be mentioned in connection with solvent extraction.

(i) Back-washing. In this technique several organic layers are combined and back-extracted into an aqueous solution of suitable composition. Since the distribution ratio for the wanted element favours the organic phase, little of this is lost; but since the distribution ratio for the contaminants is low, these are extracted back into the aqueous phase to a large extent.

(ii) Masking agents. With this technique, cyanide, tartrate, citrate, fluoride, or E.D.T.A. is added to give a charged unextractable species, in solvent extraction of uncharged chelates. To make this process clear, consider a mixture of nickel and cobalt ions. Addition of cyanide ion will give the species $Ni(CN)_4^{2-}$ and $Co(CN)_6^{3-}$. If hydrogen peroxide is now added, followed by dimethylglyoxime (DMG) we form a nickel DMG chelate, while the cobalticyanide is unattacked. These two quite different species, one uncharged, the other charged, can then be quite simply separated by extraction (see p. 78).

Chromatography is a slower partition technique. It involves a solid adsorbent as stationary phase packed into a column, and over this flows a solution with two or more components. These migrate down the column at different rates and are separated. Common adsorbents are alumina, silica gel, carbon, cellulose, and ion-exchange resins. The

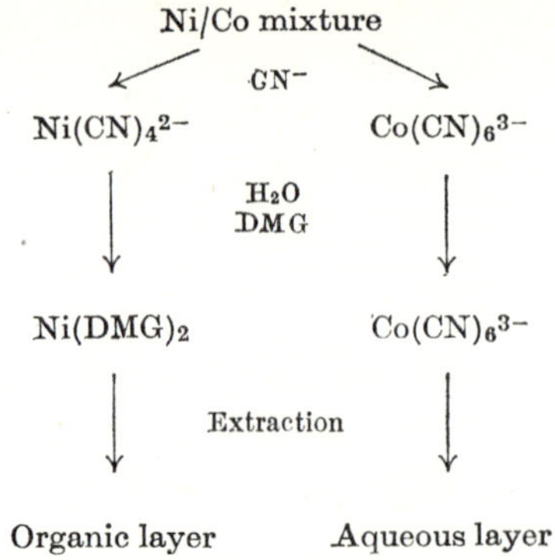

eluting agent must be less strongly adsorbed on the column than the ions we wish to separate. In radiochemistry, ion-exchange resins find the greatest use, in particular mono-functional resins based on a three-dimensional polymer network with the operative radicals, —SO_3H, —COOH, —PO_3H_2, —OH (see example on facing page).

The basic equilibria involved in the separation of two cations are,

$$HR + Na^+ \rightleftharpoons NaR + H^+$$

$$HR + Cs^+ \rightleftharpoons CsR + H^+$$

When a liquid containing these two ions travels over the stationary resin, the equilibria described will tend to hold back the ions on the resin, but they will be held back to different extents, and they emerge separated. The chief disadvantage of this method is the time involved.

Precipitation reactions are extensively used in radiochemical separation. We shall consider the effects of the precipitation of *AB*, from A^{++} and $B^=$, on another cation in the solution, C^{++}.

(i) If we precipitate *AB*, and it happens that *CB* is insoluble, then *C* is both absorbed and adsorbed on *AB*, e.g. Bi^{3+} is carried on barium carbonate, but not on barium sulphate, because bismuth carbonate is insoluble, while bismuth sulphate is soluble.

(ii) C^{++} ions can undergo isomorphous replacement in *AB*, e.g. Ra^{++} in barium sulphate.

(iii) C^{++} ions may be surface adsorbed on negatively charged aggregates of *AB*.

$$\begin{array}{c} B^= \\ A\,B\,A\,B\,A \\ {}^{++}C\;{}^=B\,A\,B\,A\,B\,A\,B\,A\,B^=\;C^{++} \\ A\,B\,A\,B\,A \\ B_= \end{array}$$

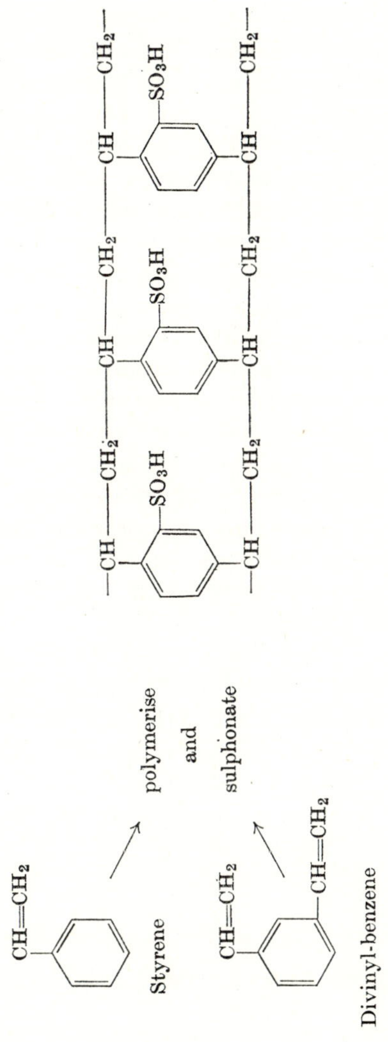
Styrene
Divinyl-benzene
polymerise
and
sulphonate

(iv) C^{++} may be incorporated in AB, even though isomorphous replacement is unlikely. When this occurs we have formed anomalous mixed crystals, e.g. rhombic plumbous chloride carried in monoclinic barium chloride dihydrate crystals.

(v) Internal adsorption can occur to give rise to unexpected contaminations.

Precipitation is a rapid method, and of universal applicability. It suffers from the disadvantage, as we have seen, that most precipitates adsorb foreign elements to some extent. The separation of solution from precipitate can be accomplished in two ways—filtration, which is slow and thorough, but contaminates large amounts of glassware, and centrifugation which is fast and contaminates much less apparatus.

An important concept when using precipitation reactions is that of scavenging, a process which is used in the removal of unwanted impurities. The best scavengers are gelatinous precipitates with a large surface area suitable for adsorption. The ions which are adsorbed depend on the charge on the surface of the precipitate; e.g. it is generally positive for silver halides, and negative for sulphides. It should be noted that univalent ions are poorly adsorbed and cannot be efficiently removed by scavenging; on the other hand, polyvalent ions are efficiently removed.

Finally we shall briefly consider electrodeposition. Since the range of voltage over which any given metal is deposited overlaps the range of other metals, electrodeposition has not been used as a primary method after activation, but is sometimes useful for final preparation of a pure source for subsequent counting.

In conclusion, we can draw up a few general rules for devising radiochemical experiments. We must devise a fast efficient first step to remove the wanted nuclide from the bulk of the sample. Poor decontamination can be permitted at this stage. Appropriate hold-back carriers for all interfering elements should be added, and at a later stage, scavengers introduced to remove all extraneous activity. Also required is a step specific for the nuclide in question.

Suggested Reading

1. Overman, R. T. and Clark, H. M. (1960). *Radioisotope Techniques.* London: McGraw-Hill.
2. Johnson, N. R., Eichler, E. and O'Kelly, G. D. (1963). "Nuclear Chemistry," *Techniques of Inorganic Chemistry*, Vol. II. London: Interscience.
3. Smales, A. A. and Wager, L. R. (1960). *Methods in Geochemistry.* London: Interscience.
4. Bowen, H. J. M. and Gibbons, D. (1963). *Radioactivation Analysis.* Oxford: Clarendon.

5. Haissinsky, M. (1964). *Nuclear Chemistry and its Applications*. Translated by D. G. Tuck. Addison Wesley.
6. Morrison, G. H. and Frieser, H. (1957). *Solvent Extraction in Analytical Chemistry*. New York: Wiley.
7. Friedlander, G., Kennedy, J. W. and Miller, J. M. (1965). *Nuclear and Radiochemistry*, 2nd ed. London: Wiley.
8. *Radiochemistry of the Elements*, National Academy of Sciences, National Research Council, Nuclear Science Series, U.S.A.E.C. (NAS-NS 3001, etc.) Washington D.C., 1960 onwards.

Chapter 13

ERRORS IN ACTIVATION ANALYSIS[1]

D. GIBBONS

United Kingdom Atomic Energy Authority,
Wantage Research Laboratory, Wantage, Berks., England

Whether activation analysis or chemical analysis is being used, errors may arise through the process of sampling. However, because of the very high sensitivity of activation analysis, and the limitations of space during irradiation, smaller samples are used. Thus any inhomogeneities in the sample are enhanced, and the relative contamination is greater. One of the major advantages of activation analysis is that the sample cannot be contaminated after the irradiation period (or, rather, that any non-radioactive post-irradiation contamination will not affect the result). There are still, however, a large number of possible ways of contaminating the sample during its preparation, for example by dust from the atmosphere, or by the acts of cutting or grinding it. Silica knives and plastic scalpels are often used for cutting biological tissues. When a biological sample is ashed or dried before irradiation elements may be lost into the atmosphere or by adsorption onto the silica vessel. Thus it may not be good practice to use a new silica vessel for each ashing operation—possibly the use of an old vessel may reduce the adsorption loss from the sample. Ion exchange resins, when they are used, may contain unwanted trace elements which could contaminate; they should therefore be separately checked for such traces.

For transfer to the reactor or neutron generator, the samples are placed in small containers usually of silica, plastic or aluminium. These containers seldom have perfectly clean surfaces, and the sample may pick up contaminants when being put into the container or when being removed for counting. In the latter case, although contamination occurs after activation, the contaminant has also been irradiated and so it is still a source of error. In cases when the sample is not removed from the container before counting, the surface cleanliness of the container is equally important, but trace elements in the material of the container are also contaminants. Clearly the standard which is usually irradiated along with the sample must be in exactly the same type of container, both from the point of view of contamination, and because the container wall reduces the neutron flux by attenuation. A silica wall 0·2 mm thick reduces the flux by as much as 5%, although even 2 cm aluminium has very little effect.

When aqueous solutions are irradiated, the water in the sample increases the thermal neutron flux by moderating the fast neutrons. This is obviously important if the solute is being activated by thermal neutrons. Other samples in adjacent irradiation positions may also distort the neutron flux distribution in the reactor, but the predominant flux effect is probably due to the flux gradient of the reactor itself. In the old graphite-moderated type of reactor this gradient can be less than about 1%, but it is much greater in small modern reactors (the neutron flux gradient in the DIDO reactor at Harwell is over a hundred times that of the BEPO reactor). Errors arising from this flux distribution can be diminished by reducing the size of the sample (although this increases the errors of inhomogeneity and contamination) or by ensuring that the exact orientation and position of irradiation can be reproduced for irradiation of other samples and the standard. If, however, the standard is irradiated at the same position and therefore at a different time, small random flux variations with time cannot be accounted for.

The effect of neutron flux gradients within the reactor can be minimized by rotating or "spinning" the sample during irradiation.

Self-shielding of neutrons by the sample is a further problem for elements with large neutron absorption cross-sections. The flux in the centre of the sample may then be much less than that at the surface. In the ideal case, when the standard has not only the same size and shape as the sample, but also the same concentration and distribution of elements, this flux variation would be unimportant. This ideal case of exact similarity is unattainable, but should be approached as closely as possible. It may be possible to add to the sample a trace element which can act as an internal standard. This is one way around the problem, but it assumes that both the unknown element and the standard trace are distributed in the same non-uniform way throughout the sample. In cases like this, the sample may often be powdered and spread homogeneously in an inert material (e.g. sucrose), but the powdering process introduces contamination, and the dilution reduces the accuracy of measurement.

The effects of self-shielding can be shown experimentally by using samples of various masses and plotting the specific activity induced against the mass of sample (specific activity = total activity/mass of sample). Fig. 32 shows the result obtained, compared with the result to be expected neglecting self-shielding. Fig. 33 shows the results of one such experiment in which the sample was a cylinder of constant cross-section, but varying length. For very small samples the self-shielding is not important, but after a fairly sharp decrease in activity with length the specific activity remains constant. This is because the self-shielding becomes so serious, that the ends of the cylinder have no effect, and the activation is simply proportional to the sample length.

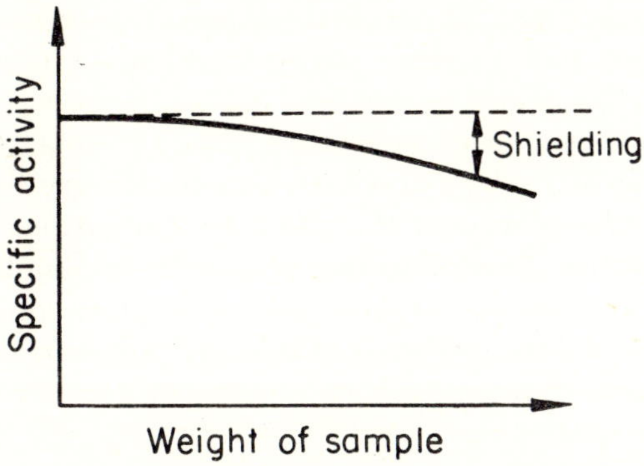

FIG. 32. Effect of self-shielding in a spherical sample.

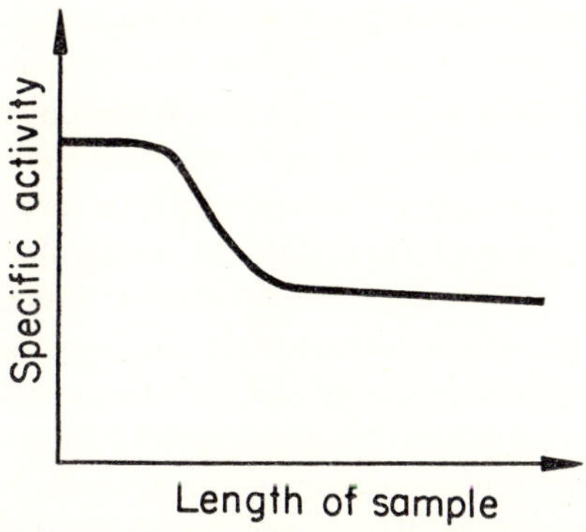

FIG. 33. Effect of self-shielding in a cylindrical sample.

Errors may arise through the occurrence of alternative neutron-induced reactions. Trace elements of atomic numbers greater by 1 or 2 than that of the element to be determined, may take part in (n,p) or (n,α) reactions respectively, and increase the concentration of that element. Similarly (n,γ) reactions on the element one place lower in the periodic table, followed by β-decay, may increase the "unknown" element concentration. It is therefore necessary to check the probable concentrations of all the neighbouring isotopes and to consider what effects they may produce.

After the irradiation the only contamination which is important is contamination by radioactive materials. It is therefore necessary that the laboratory should be clean as far as radio-isotopes are concerned. This presents little problem in laboratories using only short-lived isotopes; if isotopes with long half-lives are used in the laboratory, elaborate cleaning procedures may be necessary if good sensitivity is required. An acidified solution of the carrier element (if radio-chemical separations are carried out) is sometimes used.

If, after a radiochemical separation, the final precipitation does not give the pure elemental form, there is a possibility of the compound being non-stoichiometric and the yield determinations will be in error.

There is a further possible source of error that should be mentioned. Some samples, for example rocks and meteorites, contain elements in unusual isotopic ratios. To determine the concentration of such an element the standard must have the same isotopic ratio. In fact activation analysis has often been used for determining the isotopic ratios of elements, because it measures the amount of a particular isotope rather than a particular element.

Finally, many errors arise through the measurement of the activity, the γ-ray spectrometry, and the counting statistics. These errors are usually unavoidable, but may sometimes be diminished by the use of a computer which can cope with every available piece of data.

As in β-activity, scattering errors can occur when γ-rays are being counted. For this reason also, the necessity of having both sample and standard of similar size and shape should be emphasized.

The measurement of a decay curve is often not accurate enough to show up an interference of 10% from another isotope, especially when made on a logarithmic plot and when the half-lives are similar.

The sample and the standard cannot both be measured simultaneously on the same detector, and as all crystal detectors differ slightly in their operation parameters, it is often wise to make alternate counts of the sample and the standard. Corrections must then be made, using the experimentally determined half-life (rather than the value cited in the literature, if they differ)—this may reduce any systematic errors. All counting measurements, even with ratemeters, give the average count over a period of time. If one assumes that the average count occurred at the mid-point of the count period, then it can be shown that the error involved in this assumption is less than 1%, provided that the count period is less than the activity half-life.

The errors involved in γ-ray spectrometry due to interferences are many and varied, and have been discussed elsewhere. The measurement of counting time must be inaccurate to a certain extent. It is usual to count for a pre-set real time and to use the live-timer to calculate the count rate. If for example the live-timer counts in units of 1 sec, the real time of the count must be such that at least 100 sec of live-time occur in order to reduce the error below 1%.

Counting statistics set a theoretical limit to accuracy. If a simple count is made, the standard deviation $\sigma = \sqrt{N}$ where N is the total number of counts. The situation is more complicated in γ-ray spectrometry because of the methods used for the measurement of peak area which are at best only good approximations.

It has been shown above that activation analysis, while remaining a very sensitive technique, is also subject to a large number of errors. With the maximum amount of care, as has been taken for example in the estimation of the uranium contents of ores, an accuracy of 1–2% may be claimed. As a rough guide to the accuracies now normally accepted, radiochemical methods give 2–5% error and γ-ray spectrometry gives 5–15% error. This latter figure is at present being reduced as circuits become transistorized and are made increasingly stable.

Reference

1. Bowen, H. J. M. and Gibbons, D. (1963). *Radioactivation Analysis*. Oxford: Clarendon Press.

Chapter 14

FACTORS AFFECTING THE ACCURACY OF ASSAY

H. W. WILSON

Scottish Research Reactor Centre,
East Kilbride, Glasgow, Scotland

There are four major factors detrimentally affecting the accuracy of measurements of radioactivity. These will be briefly reviewed in turn.

A. Statistical Fluctuations in the Number of Counts

Because radioactive disintegration is fundamentally a random process, it is impossible to state with any precision the number of disintegration counts which will occur in any time interval. However if the half-life of the material is long compared with the time taken for the experiment, so that decay of the substance may be neglected, we may calculate the average number of counts per minute. Let this be called N. Suppose we now measure the actual number of counts in a given time interval a large number of times and plot a histogram of the frequency of occurrence against the number of counts (see Fig. 34). This shows a Gaussian distribution curve with a peak at N, the mean value. The narrower the peak, the more accurately may the mean value N be determined, and the shape of the curve is specified by the standard deviation σ where

$$\sigma = \sqrt{N}$$

It is found that 68·3% of the values of N counts lie between $(N+\sigma)$ and $(N-\sigma)$.

Another parameter by which the curve width may be specified is the probable error, which is defined so that 50% of the total counts are in the range between $(N+\text{P.E.})$ and $(N-\text{P.E.})$, and it is equal to $0{\cdot}675\sigma$.

For example, suppose we are to draw the exponential decay curve of a material of half life 10 hours. We may take counts of 1 min duration every half hour, and for each count calculate the probable error as equal to $0{\cdot}675\sqrt{N}$. These counts may then be plotted on a graph (see Fig. 35) and the probable error for each count shown as a vertical line through the point. We should then expect the smooth exponential curve to pass through approximately half of the lines.

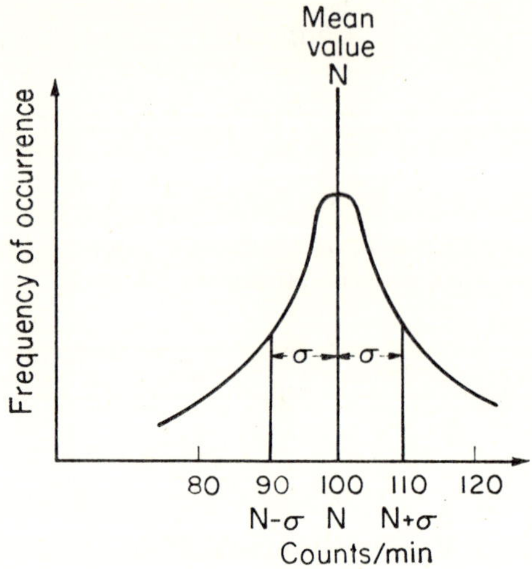

FIG. 34. Histogram of radioactivity counts.

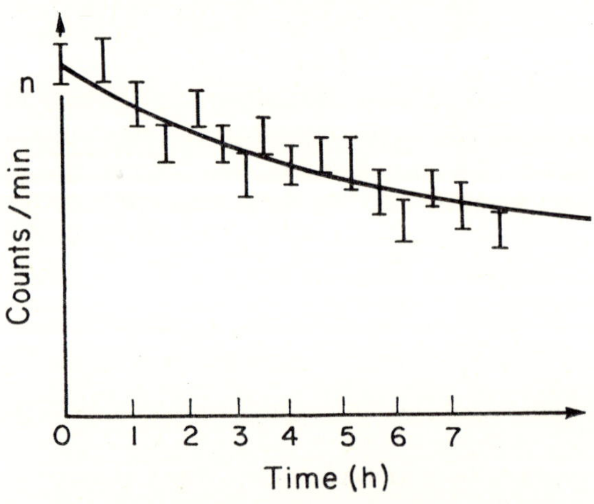

FIG. 35. Probable errors of individual counts.

The standard deviation on a single count may be decreased by taking counts over a larger interval of time, as shown by the following example.

Suppose $N = 100$ counts per minute.

If a count is made for 1 min, the standard deviation σ which is $\sqrt{N}$, is

10. We may therefore write $N = 100 \pm 10$. For a 10-min count (totalling 1000) $\sigma = \sqrt{N} = 31{\cdot}6$.

Therefore in 10 min the count is $1000 \pm 31{\cdot}6$ and the count rate may be written $N = 100 \pm 3{\cdot}16$.

By increasing the time over which a measurement is taken, the true count rate may therefore be more accurately assessed. In cases of short half life isotopes, however, the method is limited because the time taken for the count must be short compared with the half-life.

B. The Effect of Background and Counter Efficiency

When two results each with a standard deviation are to be arithmetically combined, the following rules hold:

If the results are $A \pm a$, and $B \pm b$, their sum is

$$(A+B) \pm \sqrt{(a^2+b^2)}$$

their difference is

$$(A-B) \pm \sqrt{(a^2+b^2)}$$

their product is

$$(A \times B) \pm AB\sqrt{\left(\frac{a^2}{A^2}+\frac{b^2}{B^2}\right)}$$

their ratio is

$$\frac{(A)}{(B)} \pm \frac{A}{B}\sqrt{\left(\frac{a^2}{A^2}+\frac{b^2}{B^2}\right)}$$

Of these four rules, only the first two will be required in the present discussion; the others are included for completeness. Now there exists an ever-present background radiation due to various sources, e.g. cosmic rays and radioactivity in and around the laboratory. The number of background counts depends on the shielding precautions and has to be taken into account. Let us call the background count rate N_b counts per minute, and the source count rate in which we are interested N_s. Then the total count rate will be N_t, where

$$N_t = N_s + N_b$$

As we wish to find the source count rate this equation is used in the form

$$N_s = N_t - N_b$$

The background count rate is determined by observing the number of counts, N_B, over a fairly long time interval t_B minutes, when the source is absent.

Then

$$N_b = \frac{N_B}{t_B} \pm \frac{\sqrt{N_B}}{t_B}$$

The total number of counts N_T, with the source in place, in a time t_T, is then measured and we obtain

$$N_t = \frac{N_T}{t_T} \pm \frac{\sqrt{N_T}}{t_T}$$

We may now subtract the first of these two equations from the second by means of the subtraction rule above, to get

$$\text{Source rate } N_s = N_t - N_b = \left[\left(\frac{N_T}{t_T} - \frac{N_B}{t_B}\right)\right] \pm \sqrt{\left(\frac{N_T}{t_T^2} + \frac{N_B}{t_B^2}\right)}$$

or

$$N_s = \left[\frac{N_T}{t_T} - \frac{N_B}{t_B}\right] \pm \sqrt{\left(\frac{N_t}{t_T} + \frac{N_b}{t_B}\right)}$$

This formula gives the source rate and its standard deviation. Clearly if the background rate can be reduced by more effective shielding the standard deviation of the result will also be reduced, and this is one way of achieving greater accuracy. Greater accuracy may also be achieved by using a more efficient counter. This may best be illustrated by an example.

If we have a 20% efficient counter which gives the following counts: background $N_B = 45$ in $t_B = 15$ min ($N_b = 3$ c.p.m.) and total count $N_T = 180$ in $t_T = 10$ min ($N_t = 18$ c.p.m.) then we find

$$N_s = 15 \pm \sqrt{\left(\frac{18}{10} + \frac{3}{15}\right)} = 15 \pm 1{\cdot}4 \text{ c.p.m.}$$

this represents an error of about 10%.

Let us suppose we now have a counter which is 100% efficient so that we could count five times as many particles as previously. We also assume that the background goes up five times although the increase would not necessarily be so great. Then clearly $N_B = 225$ in $t_B = 15$ min ($N_b = 15$ c.p.m.), $N_T = 900$ in $t_B = 10$ min ($N_t = 90$ c.p.m.) and we find

$$N_s = 75 \pm \sqrt{\left(\frac{90}{10} + \frac{15}{15}\right)} = 75 \pm 3{\cdot}1 \text{ c.p.m.}$$

This represents an error of only 4%. We may therefore obtain a better result by using a more efficient counter. It is because the efficiency of a Geiger counter for γ-rays is only of the order of a few per cent that a scintillation counter is preferred for analysis of this radiation.

C. Effect of Dead Time

After a particle has passed through a Geiger counter and the avalanche of ions has commenced, there is a short time in which a second particle will not be recorded as a separate pulse. This is termed the "dead-time", τ, and is of the order of 100 μs to 400 μs for a Geiger counter. Scintillation counters have a much shorter dead time, of the order of a few microseconds, but the electronic pulse height analyser introduces a longer dead-time which is superimposed upon this. A simple calculation corrects for the time during which the counter is inoperative.

Let N_t be the true count rate (to be determined) and let N_m be the measured count rate, which is less than the true count rate because of the dead-time τ.

Then

$$N_t = \frac{N_m}{1-N_m\tau}$$

Because the dead-time of the counter is not absolutely constant, a constant dead-time called the paralysis time is often introduced into the electronics of a Geiger counting system. This must be larger than τ, and is usually of the order of 400 μs.

D. Counter Geometry, Absorption and Back Scattering

The case of the β-ray source and a gas counter (Geiger or proportional) will be considered first. The efficiency of the counting system is highly dependent on the geometrical arrangement of the source and counter. This efficiency factor should be well defined—it is typically of the order of 15–20%. It is normal to find this efficiency (or otherwise eliminate the necessity of finding it) by carrying out a calibration counting experiment with a standard source.

The count is further reduced by the absorption of β-rays in the counter window, in the air between the source and the counter, and indeed in the source itself. This last is called self-absorption. Absorption loss depends upon the energy of the β-rays and, whilst negligible for high energy electrons of the order of 2 MeV, say, it is highly significant for low energy β-rays such as the 67 keV maximum energy electrons from ^{63}Ni. The effect of self-absorption may be shown experimentally by using sources of varying thickness and plotting the count rate against the thickness or weight of the source (Fig. 36). The self-absorption is seen as the deviation of the resulting curve from linearity.

Radiation is scattered through large angles by the source itself, by the source support and the walls of the shielding. This back scattering increases the count rate by a small amount which is difficult to calculate in advance.

In order to take account of these above factors the procedure often adopted in activation analysis is to irradiate a known sample of the element along with the specimen. The radioactivity induced in both

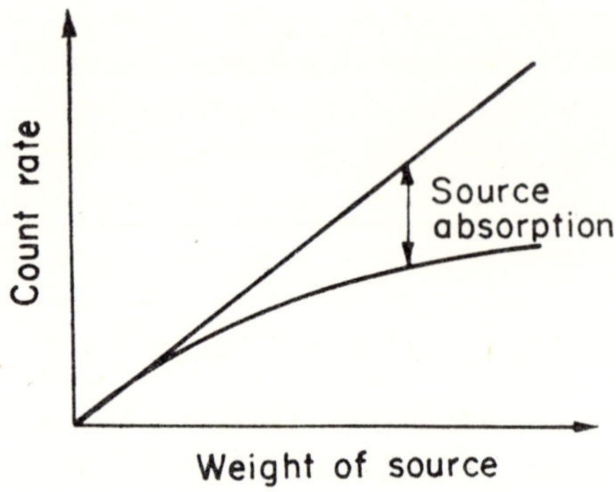

FIG. 36. Self-absorption of β-rays.

the sample and the specimen is then measured under conditions as precisely similar as possible. Effects such as absorption, counter geometry, and back scattering, are then automatically taken care of, because they affect both measurements in the same way.

In scintillation counting, geometry of the source, energy of the radiation and scattering from the shielding all have important effects on the efficiency. Self absorption is usually not significant. Again standard sources in a standard geometry are usually used to allow for these effects, although tables of efficiencies are also available for different crystal sizes, geometries and energies of radiation. With a few standard sources of different energies an efficiency calibration curve can be drawn and the efficiency for other energies found by interpolation.

REFERENCES

1. Siegbahn, K. (ed.) (1965). Chapt. 7. *Alpha-, Beta- and Gamma-Ray Spectroscopy.*
2. *A Manual of Radioactivity Procedures* (1961). N.B.S. Handbook 80, pp. 25–29, etc.
3. Friedlander, G., Kennedy, J. W. and Miller, J. M. (1965). *Nuclear and Radiochemistry*, 2nd ed. London: Wiley.

Chapter 15

COMPUTER METHODS IN ACTIVATION ANALYSIS

D. GIBBONS

United Kingdom Atomic Energy Authority,
Wantage Research Laboratory, Wantage, Berks., England

Introduction

With the development of modern methods of radionuclide measurement, the ability to record vast amounts of data reliably and automatically has led to the use of computers to interpret and evaluate that data.

Historically, the first computer programme to be used specifically for activation analysis was published only five years ago, although related problems in nuclear physics had been investigated by several laboratories prior to that. By far the major effort appears to have been in the field of γ-ray spectrum evaluation. Activation analysis with γ-ray spectrometry is developing as a rapid (and in some cases automatic) method of analysis. This development should not, however, be allowed to overshadow the other applications of computers in activation analysis. When radiochemical methods are used, the computer can be of value in interpreting decay curves, in comparing beta absorption curves, in making dead-time corrections, and in other parts of even the simplest activation calculation, particularly if many samples are to be analysed. In the general field too, the computer can be used for the optimization of irradiation, decay, and counting times.

The chief advantages of the modern computer are its high speed and versatility, which enable full use to be made of all available data by employing complex mathematical techniques which would involve a prohibitive time in manual calculations. It cannot, however, do anything that a human calculator could not do given adequate time, and in any event it cannot produce an answer any better than the counting statistics will allow. But, by making efficient use of all available data, it can give the most precise interpretation that can possibly be made.

Probably the most severe limitation of the use of a computer is that it can only do what it is told to do. Thus the human element returns in choosing or developing a correct mathematical model for solution. If this is the wrong one, but correctly programmed, then the computer will happily produce the wrong results.

Optimization Programmes

Isenhour and Morrison[1] have published a programme for the optimization of irradiation and decay times based on a modified activation equation:

$$A = INF\sigma(1-e^{-\lambda T_I})\, e^{-\lambda T_D} \tag{1}$$

where A is usable activity,

I is fraction of events in energy range measured,
N is number of target atoms,
F is flux,
σ is cross section,
T_I is irradiation time,
λ is decay constant,
T_D is decay time.

The optimum selectivity is based on the ratio R_j where

$$R_j = A_j \Big/ \sum_{i=1}^{n} A_i \tag{2}$$

However, evaluation of the maximum of (2) requires the solution of the two equations:

$$\sum_{i=1}^{n} K_i(\lambda_i - \lambda_j)\,(1-e^{-\lambda_i T_I})\,(e^{-\lambda_i T_D}) = 0 \tag{3}$$

and

$$\sum_{i=1}^{n} K_i[\lambda_j\, e^{-\lambda_j T_I}(1-e^{-\lambda_i T_I}) - \lambda_i\, e^{-\lambda_i T_I}(1-e^{-\lambda_j T_I})]\, e^{-\lambda_i T_D} = 0 \tag{4}$$

where

$$K_i = I_i N_i F_i \sigma_i \tag{5}$$

Since these equations cannot be solved directly, the method of successive approximations must be used, maximizing with respect to T_I and T_D in turn until no further significant increase in R_j is produced. This is achieved by using the criteria that if x_a is an approximate solution of $f(x) = 0$, then $x = x_a + \Delta x$ where $\Delta x = -f(x_a)/f'(x_a)$ is a better solution. To start the optimization process, a matrix of R_j values is computed at intervals over the entire range of T_I and T_D values, and the values giving the largest R_j are used as the initial trial values. This approach could well be extended to the well known problem of optimization of distribution of available counting time between sample and background, and to corrections where counting times are long compared with a half-life, and to live-time corrections when a "background" species decays significantly during the counting period.

Decay Curve Analysis

The analyses of multi-component decay curves fall into two categories: (*a*) where the number of components and their half-lives are known, and (*b*) where no information is available. In case (*a*) the total activity at time t is given as a continuous function by

$$A_t = \sum_{j=1}^{n} x_j \, e^{-\lambda_j t} \tag{6}$$

and the individual points of the decay curve by

$$C_i/T_i = \sum_{j=1}^{n} x_j \, e^{-\lambda_j t_i} \tag{7}$$

Equation (7) is a good approximation for (6) if T_i, the counting period, is less than one half-life.

This set of points is fitted to the experimental data, finding the best values of x_j by the method of least squares. This principle requires that each data point be equally reliable; they are therefore weighted as follows:

$$Y_i = A_i/\sigma_i = \sum_{j=1}^{n} (x_j \, e^{-\lambda_j t_i})/\sigma_i \tag{8}$$

This can be represented in matrix notation as

$$(Y) = [A](x) \tag{9}$$

The solution of these equations by the method of least squares gives

$$[A^T](Y) = [A^T][A](x) \tag{10}$$

or

$$(x) = ([A^T][A])^{-1}([A^T](Y)) \tag{11}$$

where $[A^T]$ is the transpose matrix of $[A]$. The inversion of the matrix $[A^T][A]$ involves some figure loss, but many alternative methods of matrix inversion have been recommended. One method used in several publications[2] proceeds by orthogonalizing $[A]$ by multiplying by an appropriate matrix $[U]$ to give $[B]$.

i.e.
$$[A][U] = [B] \tag{12}$$

Then

$$[U^T][A^T][A][U][U]^{-1}(x) = [U^T][A^T](Y) \tag{13}$$

or

$$[B^T][B][U]^{-1}(x) = [B^T](Y) \tag{14}$$

or

$$[D][U]^{-1}(x) = [B^T](Y) \tag{15}$$

Hence

$$(x) = [U][D]^{-1}[B^T](Y) \tag{16}$$

Since $[D] = [B^T][B]$ is a diagonal matrix its inversion involves only replacing each element by its inverse and no figure loss is involved.

When the half-lives are not known, the equations must also be solved for λ_j. However, even if the number of components is known, the method of least squares could not be applied as λ_j appears in an exponential term. An iterative process is therefore used in which approximate values x_j^a and λ_j^a are inserted and the exponential term expanded:

$$(x_j^a + \Delta x_j)\,e^{-(\lambda_j^a + \Delta\lambda_j)t_i} \simeq (x_j^a + \Delta x_j)\,e^{-\lambda_j^a t_i} - x_j^a\,\Delta\lambda_j\,t_i\,e^{-\lambda_j^a t_i} \tag{17}$$

if $\Delta\lambda_j . t_i \ll 1$ and $\Delta x_j . \Delta\lambda_j$ is neglected.

Here, equation (17) is linear in both Δx_j and $\Delta\lambda_j$ and can be solved by the method of least squares.

Fourier Transformation

Yet a further method of decay curve analysis, differing in its approach to the solution of a set of simultaneous equations, rather than just a different method of matrix inversion, involves the principle of Fourier transformation.[3] This method depends on the fact that

$$f(t) = \sum_{i=1}^{n} N_i\,e^{-\lambda_i t} = \int e^{-\lambda t}\,dh(\lambda) \tag{18}$$

and

$$f(t) = \int e^{-\lambda t}\,g(\lambda)\,d\lambda \tag{19}$$

where $h(\lambda)$ is a step function representing the count rate histogram, and $g(\lambda)$ is a sum of delta functions, each of which is zero except at a specific value of λ. A plot of $g(\lambda)$ against λ would give a line spectrum for the appropriate values of λ, except that the experimental errors broaden this into a series of peaks of position λ_i and height proportional to N_i. The function $g(\lambda)$ is obtained by Fourier transformation as the equation

$$g(e^{-Y})\,dY = [g(\lambda)/\lambda]\,d\lambda \tag{20}$$

which can apparently be evaluated numerically.

γ-Ray Spectra

By far the majority of published literature on the uses of computers in activation analysis relates to the unscrambling of γ-ray spectra. This may be achieved in a variety of ways.

(*a*) *Spectrum Stripping* in which the highest energy peak is identified, measured, and then subtracted from the spectrum. This was the basis of the first activation analysis programme, but required the collection of many spectra (since decay was used as an additional criterion of identification) and was purely an extension of the normal "graphical" method of stripping. However, it has also been used, in modified form, in more recent programmes.

Thus, Perkons, Hummel and Jervis[4] used a spectrum stripping technique, but without decay identification, to measure 10–14 elements in hair. The peaks were identified and measured, from the high energy end, by doing a least squares fit to a parabola, three points at a time, and subtracting a computed multiple of the appropriate standard. The programme corrected for over- and under-subtraction as it proceeded, and any residual peaks were pointed out as a difference spectrum at the end.

(*b*) *Simultaneous equations* may be used to solve a system of known components or one where the major components can be resolved and identified. The spectrum is divided up into an appropriate number of regions (preferably the photopeaks) and a set of simultaneous equations developed. Standard spectra of all the components are measured so that the background contributions at the position of the ith photopeak for the jth component can be calculated as a multiple F_{ij} of the main peak in the jth component. Thus, at each part of the spectrum,

$$C_i = \sum F_{ij} X_j \tag{21}$$

This set of equations can be solved for X_j, which are then compared with the standards to give the amounts of the elements present in the normal way.[5]

(*c*) *Least squares fitting.*[6] By extending the above model to every point in the spectrum, and including an error term Z_i, one obtains as many equations as channels, and the sum of Z_i^2 can be minimized in the normal way. Thus,

$$\sum_{j=1}^{m} K_j \sum_{i=1}^{n} A_{ik} A_{ij} = \sum_{i=1}^{n} A_{ik} C_i \tag{22}$$

where K_j is the ratio of the jth component to its standard; A_{ij} is the count *rate* in channel i of the jth standard; C_i is the count in the ith channel of the sample; and k is the equation index varying from 1–m.

(*d*) *Vector Analysis.* This considers the unknown spectrum as an n-dimensional vector (for n channels) and is represented as the sum of multiples of m standard n-dimensional vectors[7]. Since it can be shown that, mathematically, this approach does not differ from the method of least squares, it need not be considered further.

(*e*) *Iterative methods* may be applied in two ways—firstly where little or no information is available, or obtainable, with regard to the energy distribution, or secondly as a method of solving the least squares approach, when experimental errors are such that the response matrix is not singular. In either case the approach is the same as in decay curve analysis—guesses of the unknowns are first made and then corrections to these guesses are made and so on.

(*f*) *Synthetic methods.* Heath[8] has used the computer to synthesize individual γ-ray spectra, knowing details of sample energy, size, position, detector geometry, etc. and taking into account such effects as back-scatter peaks, x-ray production, β-absorption, self absorption, summation effects, and detector performance. Thus, for a given energy, the programme can predict the actual spectrum. This approach can be extended to multiple spectra, in which the spectrum is synthesized, compared with the experimental spectrum, and the synthetic model modified by the usual iterative process.

A similar method, used at Saclay[9] for mixtures of known energy, assumes (*a*) that the photopeaks are Gaussian, and (*b*) that the "background" between peaks, made up of Compton continua and low energy continua, can be represented by a sum of exponentials. Thus the counts in the ith channel are

$$C_i = \sum_{j=1}^{j_{\max}} A_j \mathrm{e}^{-B_j N_i} + \sum_{k=1}^{k_{\max}} D_k \mathrm{e}^{-E_k(N_i - N_k)^2} \tag{23}$$

where $j_{\max}$ is the maximum number of exponentials required to fit the background; $k_{\max}$ the maximum number of photopeaks; N_k the peak locations; and A and D are the intensities of backgrounds and photopeaks. An iterative process is used to minimize the Chi squared function given by:

$$\chi^2 = \sum_{i=1}^{i_{\max}} \left[\frac{(C_i - Q_i)}{\Delta Q_i}\right]^2 \tag{24}$$

where Q_i is the observed counts in the ith channel and ΔQ_i the associated standard error.

(*g*) *Errors.* One advantage of matrix methods is that in solving equations such as

$$(x) = [S^{-1}](z) \tag{25}$$

the errors of the results turn up as the diagonal of the inverse matrix, and so are directly available. Burrus[10] has shown that by the use of prior information, such as the non-negativity of concentration, or of peak identity in iterative methods, the errors can be reduced significantly.

OTHER APPLICATIONS

(1) *Spectrum smoothing.* γ-Ray spectra can be smoothed either before applying least squares fitting, or after applying certain iterative procedures, by a variety of techniques.[11] These may be illustrated by the simple approach in which the spectral points are fitted, say five at a time, to a third order polynomial, re-calculating the mid-group point by the equation

$$C_n = \frac{1}{35}[-3C_{n-2}+12C_{n-1}+17C_n+12C_{n+1}-3C_{n+2}] \tag{26}$$

A more complicated approach treats the spectrum as an electrical signal which is, in effect, passed through a "Gaussian" filter to remove unwanted "noise" by performing a convolution integral between the γ-spectrum and the filter response curve.

(2) *Peak location.* All the methods of peak location mentioned elsewhere, can of course be programmed for computer evaluation, but it must be remembered that least squares fitting can only be applied to linear functions of the data points, and not to derivatives of those points arranged to give linear functions. Thus, in using the Gaussian property of the photopeak, a least squares fit of the logarithms of the data points to a parabola, although possible, will produce the answer which minimizes the logarithms of the residuals, and not necessarily the residuals themselves. Once again we turn to an iterative procedure and guess the central value, and also the peak width (although this *could* be known from the resolution of the instrument). The function is already linear in the height of the peak, so this does not need to be guessed. The corrections to the initial guesses are computed to the required degree of accuracy and the Gaussian equation solved.

(3) *Drift control.* Where automatic drift control apparatus is not available, and where drift can be restricted manually within certain limits, corrections for drift can be made during the evaluation of the γ-spectrum. Suitable peaks in the spectrum (natural or added) are located and the deviations in their positions from those expected are determined. The changes in gain and/or zero-level can then be calculated and the γ-ray spectrum corrected.

(4) *Flux monitoring.* The effect of time dependent flux variations on

sample activity in neutron generator activation analysis may be considered as follows:[12]

If the relative flux level is measured every minute then, during the first minute, any one isotope will be activated according to:

$$(A_i)_1 = \frac{0{\cdot}602\phi_1 \sigma_i W_i}{M_i}\left[1-\exp\left(\frac{-0{\cdot}693}{t_i}\right)\right]$$

Where $(A_i)_1$ is the induced activity in the ith isotope in the first minute,
ϕ_1 is the effective flux during the first minute,
σ_i is the activation cross-section of the ith isotope,
W_i is the mass of the ith isotope present,
M_i is the atomic weight of the ith isotope,
t_i is the half life of the ith isotope.

Similarly, in the jth minute

$$(A_i)_j = \frac{0{\cdot}602\phi_j \sigma_i W_i}{M_i}\left[1-\exp\left(\frac{-0{\cdot}693}{t_i}\right)\right]$$

and so on up to a total of T_{irrad} minutes.

At the end of T_{irrad} minutes, the activity produced in the first minute will have decayed to:

$$(A_i')_1 = \frac{0{\cdot}602\phi_1 \sigma_i W_i}{M_i}\left[1-\exp\left(\frac{-0{\cdot}693}{t_i}\right)\right]\left[\exp\left(-0{\cdot}693\frac{T_{\text{irrad}}-1}{t_i}\right)\right]$$

and the activity produced in the jth minute will have decayed to:

$$(A_i')_j = \frac{0{\cdot}602\phi_j \sigma_i W_i}{M_i}\left[1-\exp\left(\frac{-0{\cdot}693}{t_i}\right)\right]\left[\exp\left(-0{\cdot}693\frac{T_{\text{irrad}}-j}{t_i}\right)\right]$$

Thus the total activity due to the ith isotope at the end of the irradiation will be:

$$(A_i)_{\text{total}} = \sum_{j=1}^{j=T_{\text{irrad}}} (A_i')_j$$

$$= \frac{0{\cdot}602\sigma_i W_i}{M_i}\left[1-\exp\left(\frac{-0{\cdot}693}{t_i}\right)\right]\sum_{j=1}^{j=T_{\text{irrad}}} \phi_j\left[\exp\left(-0{\cdot}693\frac{T_{\text{irrad}}-j}{t_i}\right)\right]$$

Hence, knowing the values of the measured flux, the value of the measured total activity and the values of the constants, W_i can be calculated. However, in activation analysis it is more usual to obtain the value of W_i by comparison of the induced activity with that induced in a known standard. Also, the measured flux values in this experiment are relative and not absolute.

If the flux had been constant during activation, the induced activity would have been

$$(A_i)_{\text{total}} = \frac{0{\cdot}602\sigma_i W_i}{M_i} . \phi_{\text{eff}}\left[1-\exp\left(-0{\cdot}693\frac{T_{\text{irrad}}}{t_i}\right)\right]$$

Hence the effective flux ϕ_{eff} may be written as

$$\phi_{\text{eff}} = \frac{\left[1-\exp\left(\frac{-0{\cdot}693}{t_i}\right)\right] . \sum_{j=1}^{j=T_{\text{irrad}}} \phi_j\left[\exp\left(-0{\cdot}693\frac{T_{\text{irrad}}-j}{t_i}\right)\right]}{\left[1-\exp\left(-0{\cdot}693\frac{T_{\text{irrad}}}{t_i}\right)\right]}$$

The unknown mass may now be calculated from the simple relationship

$$\frac{(W_i)_{\text{sample}}}{(W_i)_{\text{standard}}} = \frac{((A_i)_{\text{total}})_{\text{sample}}}{((A_i)_{\text{total}})_{\text{standard}}} . \frac{(\phi_{\text{eff}})_{\text{standard}}}{(\phi_{\text{eff}})_{\text{sample}}}$$

(5) Once counting data is in the computer memory, other procedures are advantageously performed. Thus, γ-ray spectra and decay curves, etc., can be plotted, using the high definition plotters now available with many computers, or even by programming the printer to produce an acceptable plot.

The other calculations involved in activation analysis, although trivial by comparison, are still worth programming, if only to make the process complete. Thus, corrections for irradiation time, self-shielding, decay-time, live-time, self-absorption, etc. and the final calculation of elemental concentration, could all be included.

If each result is summarized on a "file" card by the computer, then at appropriate times these can be evaluated with regard to matrices and elements analysed, concentration ranges, errors, time and cost, etc.

To summarize, computer methods are free from subjective errors; they can be used to evaluate statistically poor data; the whole of the available data can be taken into consideration; errors are readily calculated; the methods are rapid, reliable and cheap (say 1*s*. per sample).

The limitations are in the versatility and reliability of the programming, and the fact that the computer cannot do better than the errors will allow. The limited direct availability of computers is not serious, since in most countries computer services now operate. The only problem here is one of overall time.

Future

In looking to the future it is possible to visualize the optimization programme being extended to the whole analysis. Thus, the computer would first be supplied with information on the element(s) to be measured, expected concentrations, and any information concerning the matrix. It would then suggest appropriate irradiation, decay, and counting times, for reactor, accelerator or cyclotron analysis, energy settings of spectrometer, whether radiochemistry is necessary, photopeaks to be measured, and if decay curves were to be obtained. A trial run would be made and the results fed back and modified instructions obtained. A further run should then give acceptable results.

With the development of more specialized small computers, and the reduction in costs which go with increased demand, a computer designed specifically for activation analysis may one day be worthwhile. This might have the gamma spectrometer memory as a part of the computer memory to give direct data processing.

Developments in other fields could also be incorporated. Thus, automatic radiochemical separation systems have already been described, and these will become more freely available with time. The ultimate is then the fully automatic activation analysis laboratory, in which a sample can be dropped in and, if an analysis can be made at all, it will be made as rapidly as possible with whatever techniques are necessary automatically under full computer control.

References

1. Isenhour, T. L. and Morrison, G. H. (1964). *Anal. Chem.* **36**, 1089.
2. Nervik, W. E. (1963). "Applications of Computers to Nuclear and Radiochemistry," NAS-NS 3107, 9.
3. Gardner, D. G. and Gardner, J. C. (1963). "Applications of Computers to Nuclear and Radiochemistry," NAS-NS 3107, 33.
4. Perkons, A. K., Hummel, R. L. and Jervis, R. E. (1963). 2nd National Meeting Society for Applied Spectroscopy, San Diego.
5. Breen, W. M., Fite, L. E., Gibbons, D. and Wainerdi, R. E. (1961). *Trans. Amer. Nuclear Soc.* November.
6. Salmon, L. (1963). "Applications of Computers to Nuclear and Radiochemistry," NAS-NS 3107, 165.
7. Summers, D. L. and Babb, D. D. (1963). "Applications of Computers to Nuclear and Radiochemistry," NAS-NS 3107, 143.
8. Heath, R. L. (1963). "Applications of Computers to Nuclear and Radiochemistry," NAS-NS 3107, 93.
9. Freeman, R. and Mani, G. S. (1964). *Nucl. Instr. Methods*, **26**, 34.
10. Burrus, W. R. and Bogert, D. (1963). "Applications of Computers to Nuclear and Radiochemistry," NAS-NS 3107, 127.
11. Fite, L. E., Steele, E. L. and Wainerdi, R. E. (1962). Texas Engineering Experiment Station report TEES-2671-2.
12. Gibbons, D., Fite, L. E. and Wainerdi, R. E. (1962). U.S.A.E.C. report TID-7629, 221.

Chapter 16

GEOCHEMICAL AND COSMOCHEMICAL APPLICATIONS

H. J. M. BOWEN*

United Kingdom Atomic Energy Authority,
Wantage Research Laboratories, Wantage, Berks., England

Activation analysis as a technique has completely revolutionized the science of geochemistry over the past decade, altering very appreciably our ideas on the composition of the earth's crust, sea-water, meteorites, etc. The technique is of course highly sensitive, although in many cases it is possible to obtain large samples and the sensitivity itself is then not the most important feature. There is also the advantage of freedom from contamination by reagent material. Conventional analysis involves dissolving crushed rock in potent acids or in molten sodium peroxide which contaminate the sample, but contamination of rock samples after activation is unimportant. Another great advantage of activation analysis over chemical methods is that specific isotopes are detected rather than specific elements. This is essential when the analysis is used for sample dating or for determining the relative abundances of nuclides in the cosmos.

As in any analytical method, sampling presents a major problem. Even in temperate regions weathering processes extend to a depth of about 30 m, and in tropical countries this may be increased to 100 m. Contamination problems also occur. Samples are often obtained from museums, where they may at some time have been varnished or treated with copper sulphate solution, and where they have been collecting dust for years. Contamination is also likely in crushing rock samples for rapid dissolution.

It is fortunate that, after irradiation, the major constituents of rock samples do not constitute a radio-activity hazard (although the same cannot be said of sea-water). The elements Fe, Si, O and Ca do not activate significantly with short irradiation times; Al, Mg and Ti do activate, but they decay quickly, in a matter of minutes. Most of the γ-radiation from rock samples comes from Mn, Na and K.

* *Present address: Chemistry Dept., The University, Reading, Berks.*

Cosmochemistry

It is in this field that activation analysis has caused major advances in a short time. It is the only reliable way of analysing meteorites and micro-meteorites; many of the analyses for trace elements published over 15 years ago are now known to be incorrect.

Rocks which are suspected of being meteoritic can be tested with some degree of certainty by measuring the Ni:Cu ratio. Generally speaking this ratio is much higher in meteorites than in terrestrial rocks (with the obvious exception of nickel ores themselves). Meteorites are broadly divided into two classes: (*a*) chondrites or stony meteorites; and (*b*) iron meteorites. It is probable that the majority of meteorites are of class (*a*), but as these are less remarkable in appearance than the iron meteorites, they are less often found. In terrestrial iron nodules, nickel and copper are both present at very low levels, of the order of a few parts per million. In iron meteorites on the other hand, the nickel concentration often exceeds 10%, while copper is still only present at trace levels. Stony meteorites also have a rather high Ni:Cu ratio.

The origin of "tektites" has been a controversial question in the past, and the matter is still unsettled. Tektites are glassy bodies; many have been found in central Australia, sometimes deeply embedded in rock strata. They are either meteoritic or volcanic, and the latter view seems rather more probable now that the Ni:Cu ratio has been measured and found to be low.

One can calculate that the rate of deposition of meteoritic dust on the ocean bed should be quite significant, but, by using activation analysis techniques on marine sediments, the Ni:Cu ratio was found to be the same as that of normal terrestrial rocks. In fact the nickel and copper may be dissolved and dispersed in the sea-water before the micrometeorites are deposited as sediment.

In 1956 Suess and Urey [1] reported that 42 elements in chondrites had analyses of doubtful quality. By 1964, 34 of these have been studied in some detail and only eight elements are still doubtful. Bi, In, Tl and W have been found present at concentrations only 1/100th the concentrations supposed in 1956, and the abundances of Ag, Hg, Ta, Th, and U, have been shown to be lower by a factor of ten.

Generally speaking the abundances of the heavier elements in meteorites do not conflict with the Burbidge–Hoyle [10] theory of the synthesis of these elements in stars, although some discrepancies do exist. For example indium appears to be a much rarer element than it should be by any theory.

One of the earliest findings from activation analysis on meteorites was that iron meteorites appear to fall into three groups according to their gallium content, which may be of the order of 100 p.p.m., 10 p.p.m. or 1 p.p.m.[3] The Ga content seems to be related inversely to the nickel content, the third group containing the most nickel, and it is also reflected in variations of tin and germanium content.

The basic theories of the origin of meteorites suppose them to be the debris of a disintegrated planet, which would have had an iron core and a stone mantle. Some experimental work is being done on synthetic laboratory systems, to find the distribution of trace elements in a molten iron/rock model.

Geochemical Problems

Smales, Wager[9] and others have studied the behaviour of various elements in the crystallization of magma using activation analysis techniques. In solidified pools of magma the trace element composition varies from the central regions (which presumably were the last to solidify) to the outer layers. Broadly speaking, the regions of earlier solidification have compositions similar to a standard diabase sample termed W1, and the later regions are similar to the standard granite sample G1. Among the first elements to crystallize out of the molten magma are Co and Ni, while As, Ba, Cr, Mn, Mg and Ti also crystallize out fairly early. K, Ca, Cu and the rare earths are deposited later and the final fraction includes F, Rb, Ta, Tl, Th and U. These regions of late magma solidification are therefore the most likely regions for finding veins of uranium ore. There is further a large number of elements, such as Sr, Al and Na, which are not particularly concentrated in any region but appear throughout the magma. It has recently been found, through this work, that magmas from central Pacific regions (Hawaii) differ from European magmas in their elemental composition. This may be due to a difference in their depths of origin.

The above considerations apply to the primary igneous rocks. Sedimentary rocks also constitute a large field for research. These are mostly laid down by biological processes; in limestone for example, the calcium carbonate is precipitated by living organisms from sea-water. Interesting work has been done on barium : strontium ratios and caesium : rubidium ratios. While the concentrations of Ba and Sr are approximately equal in primary rocks, they are widely different in sedimentary rocks; also far more Rb than Cs is found in sedimentary rocks. Manganese nodules from the ocean bed have been analysed and found to contain large amounts of trace elements. It is possible that $Mn(OH)_2$ in the ocean acts as a scavenger as it precipitates, gathering large amounts of other elements, many of which are, as it happens, poisonous to marine life.

Hydrogeochemical Problems

The following are some of the trace elements which have been found in sea-water by activation, although most of them are present in extremely small amounts: Au, As, Ba, Cs, Mn, Rb, Sn, Sr, Ta, U. The presence of tantalum is a little surprising as it has no known soluble compounds. The concentrations found by activation analysis are in nearly all cases lower than had been previously supposed.

Difficulties are found in collection of the sea-water without losing trace elements by absorption onto the surface of the container. Plankton also creates problems—this may contain a lot of trace elements and has to be removed, possibly by centrifugation. It appears that the growth of plankton depends largely on the presence of traces of P, N, and Mn. If this is so, then trace quantities of Mn have a remarkable influence on world economy via the fishing industry.

The question of the oceanic mineral balance has also been studied, and it appears that the salt concentration at least has remained fairly constant since pre-Cambrian times. There may possibly be an accumulation of other elements in the sea, as appears likely for strontium. In sea water the Ca:Sr ratio is 50, while in igneous rocks it is about 500 and the ratio rises to 1000 for sedimentary rocks. It thus appears that the organisms responsible for laying down the sedimentary rocks remove the Ca preferentially, leading to a gradual accumulation of Sr in the ocean.

Chronogeochemistry

Where an element A* undergoes a radioactive transformation to an element B with a half life of the order of 1 aeon ($= 10^9$ years), an estimate of rock ages is possible if the concentrations of A* and B are measurable. The age of the earth is generally believed to be about 4·5 aeons. Four decay processes are commonly used in dating ancient rocks (only the initial and final nuclei of what may be a long chain are shown):

$$^{40}\mathrm{K} \xrightarrow{1{\cdot}3\ \mathrm{ae}} {}^{40}\mathrm{Ar}$$

$$^{232}\mathrm{Th} \xrightarrow{14\ \mathrm{ae}} {}^{208}\mathrm{Pb}$$

$$^{235}\mathrm{U} \xrightarrow{0{\cdot}7\ \mathrm{ae}} {}^{207}\mathrm{Pb}$$

$$^{238}\mathrm{U} \xrightarrow{4{\cdot}5\ \mathrm{ae}} {}^{206}\mathrm{Pb}$$

The Pb isotopes are commonly detected by mass spectrometry, although ^{208}Pb takes part in an (n,γ) reaction. Th and U are detectable through $(n,\text{fission})$ reactions. Iron meteorites have been found to be about 4·7 aeons old, mainly through the K-Ar decay. Two other long-lived nuclides may be mentioned, although the half-lives are perhaps too long to be useful;. the nuclides are also rather rare. These are:

$$^{176}\text{Lu} \xrightarrow{30\text{ ae}} {}^{176}\text{Yb}$$

$$^{187}\text{Re} \xrightarrow{50\text{ ae}} {}^{187}\text{Os}$$

All four nuclides are detectable by (n,γ) activation.

Some interest has also been shown in extinct radionuclides, which, although they may have once had fairly high abundances, have half-lives short compared with the age of the earth and have now decayed to negligible amounts. For example ^{129}I decays to ^{129}Xe with a half life of about 0·1 aeons or less. It is possible to measure the increasing abundance of ^{129}Xe and extrapolate back in time to estimate the original abundance of ^{129}I. Such data are of course extremely useful to astrophysicists studying the synthesis of the elements.

References

1. Suess, H. E. and Urey, H. C. (1956). "The abundances of the elements," *Rev. Mod. Phys.* **28**, 53.
2. Ehmann, W. D. (1961). *J. Chem. Educ.* **38**, 53.
3. Brown, H. and Goldberg, E. (1949). *Science*, **109**, 347.
4. Goldberg, E., Uchiyama, A. and Brown, H. (1951). *Geochem. Cosmochim. Acta* **2**, 1.
5. Rankama, K. (1954). *Isotope Geology.* Pergamon Press.
6. Rankama, K. (1961). *Progress in Isotope Geology.* Pergamon Press.
7. Green, J. (1959). "Geochemical Table of the Elements," *Bull. Geol. Soc. Amer.* **70**, 127.
8. Bowen, H. J. M. and Gibbons, D. (1963). *Radioactivation analysis.* Oxford: Clarendon Press.
9. Smales, A. A. and Wager, L. R. (1960). *Methods in geochemistry.* Interscience.
10. Burbidge, E. M., Burbidge, G. R., Fowler, W. A. and Hoyle, F. (1957). "Synthesis of the elements in stars," *Rev. Mod. Phys.* **29**, 547.

Chapter 17

OXYGEN IN METALS

G. I. CRAWFORD

Natural Philosophy Department, The University, Glasgow, W.2, Scotland

In the manufacture of steel, oxygen is added to molten iron to burn out the impurities it contains. However oxygen itself is an impurity, and so there is a limit to the amount which can be added without disadvantage. It is necessary to add oxygen until the total impurity level is at a minimum, and if possible to keep the oxygen content at this level. This requires a process of rapid analysis of the oxygen content in the steel, so that the oxidation process can be checked during the operation. Activation analysis can be automated to such an extent that this is possible.

The most important nuclear reaction for oxygen determination is

$$^{16}O(n, p)\,^{16}N$$

$$^{16}N \xrightarrow{\beta^-} {}^{16}O^* \xrightarrow{\gamma} {}^{16}O$$

The β-emitting nitrogen isotope has a half-life of 7·3 sec. An approximate decay scheme for the reaction is shown in Fig. 37. from which it may be

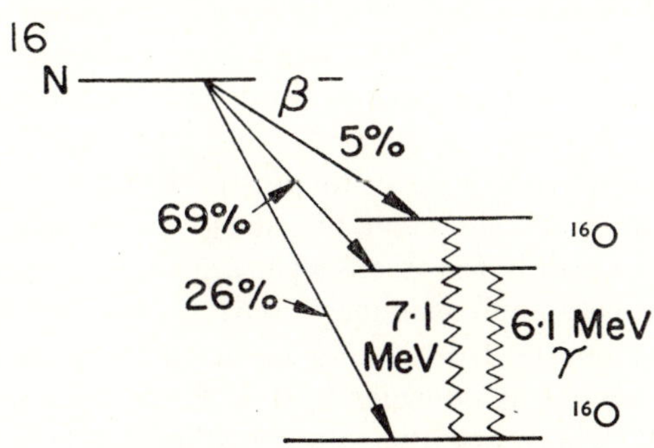

FIG. 37. Nitrogen-16 decay scheme.

seen that 7·1 MeV and 6·1 MeV γ-rays are emitted. The cross-section for the reaction is about 90 millibarns for neutrons of about 14 MeV.

The mathematics of the activation process, and the effects of irradiation time t_1, the time t_2 between the end of the irradiation and the start of counting, and the counting time t_3, may be summarized by the equation:

$$C = \frac{mN\sigma\phi}{A}(1-\mathrm{e}^{-\lambda t_1})(\mathrm{e}^{-\lambda t_2})(1-\mathrm{e}^{-\lambda t_3})E$$

In this equation

C = number of counts recorded by γ-ray detector
m = amount of oxygen in the sample (g)
N = Avogadro's number
A = atomic weight of oxygen
σ = reaction cross-section
ϕ = neutron flux
λ = decay constant of activity
E = efficiency of detection

The two increasing exponential functions, namely $(1-\mathrm{e}^{-\lambda t_1})$ and $(1-\mathrm{e}^{-\lambda t_3})$, increase to a value of about 0·98 when the irradiation time and the count time have been made equal to four half lives. This is the practical limit beyond which it is not profitable to go. For the 7·3-sec nitrogen decay this means that t_1 and t_3 should both be about 30 sec.

The decaying exponential, $(\mathrm{e}^{-\lambda t_2})$ should also be made as large as possible, i.e. t_2 should be as small as possible. The shortest practicable time for transfer of the sample from the neutron irradiation position to the counting position is about 2 sec (although for special studies this has been reduced to 0·1 sec). The total time required for the analysis, $(t_1+t_2+t_3)$, is therefore about 1 min.

The spectrum of the γ-radiation from an activated sample is shown (rather simplified) in Fig. 38. The three large peaks are the photoelectric peak of the 6·1 MeV γ-radiation, and the two "escape peaks" of this radiation at energies 0·51 MeV and 1·02 MeV less. To see how these occur, suppose that a γ-ray forms an electron-positron pair in the scintillation crystal. All the electron energy is absorbed in the crystal, but the positron annihilates itself with another electron, creating two γ-rays in the process each of energy 0·51 MeV. Either one or both of these γ-rays may escape from the crystal. In these cases a peak in the γ-ray spectrum occurs at energies 0·51 MeV or 1·02 MeV less than the original 6·1 MeV. There are also small peaks at 7·1 MeV, and 6·6 MeV due to the alternative transition in $^{16}O^*$ as seen in Fig. 37.

A discriminator or single channel analyser may be set to record only those counts in the band containing the oxygen photopeak. The γ-rays are, however, of fairly high energy and there is little interference, and so

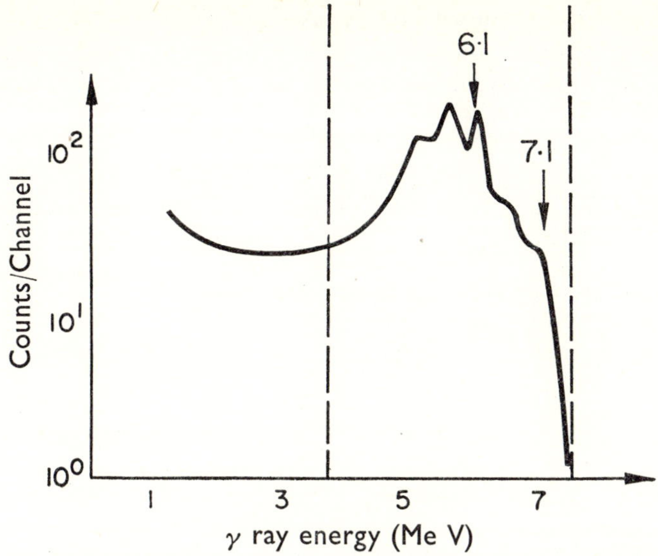

FIG. 38. γ-Ray spectrum from activated sample.

the count rate may be improved by setting the low energy channel limit at 3·75 MeV. An upper channel limit of 7·5 MeV cuts out interference from cosmic rays, which give detector pulses similar to γ-rays of about 10 MeV. Unfortunately there are two reactions which could cause interference. These are:

$$^{15}N(n,p)\,^{15}C, \qquad t_{1/2} = 2 \text{ sec}, \qquad E_\gamma = 5 \text{ MeV}$$

$$^{19}F(n,\alpha)\,^{16}N$$

Now the ^{15}N isotope only constitutes about 0·36% of natural nitrogen, and the short half-life also helps to reduce the interference of this reaction. The second reaction could be more serious, because ^{16}N is the isotope produced in the oxygen activation, and there is no way of easily checking whether the activity was produced by oxygen or fluorine. The fluorine content of steel can be checked by carrying out an activation based on

$$^{19}F(n,p)\,^{19}O$$

and it is found that the fluorine content of steels is negligible. In practice these two reactions can be ignored.

Fig. 39 shows a schematic layout of the activation system. Many types of neutron generator are now commercially available supplying fluxes of about 10^9 or even 10^{10} neutrons/cm^2/sec, but the increase in flux (and hence in sensitivity) must be balanced against an increase in cost. The γ-detection system consists of two 3 in. $\times$ 3 in. NaI crystals which provide 10% resolution. Here again the cost increase of larger crystals

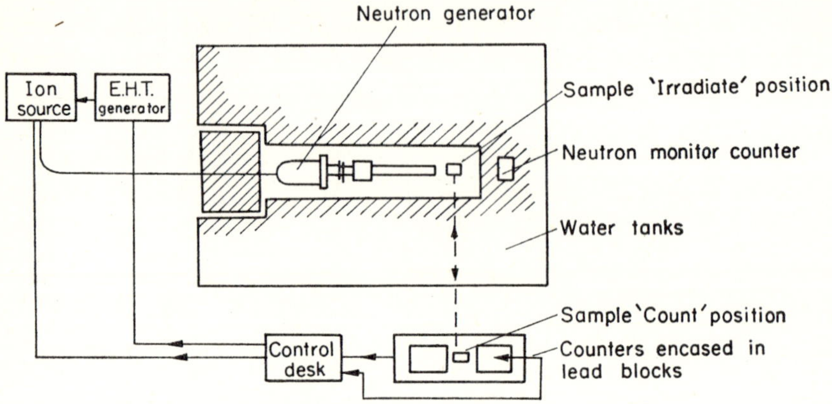

FIG. 39. Schematic arrangement of activation analysis equipment.

and 8% resolution crystals outweighs their advantages for this application. In order to set up the spectrometer it would be convenient to have a 7·5 MeV source and a 3·75 MeV source. These are not available, but a mixture of ^{60}Co and ^{137}Cs gives γ-ray lines at 1·33 MeV and 0·66 MeV which may be used instead. The gain of the amplifier is easily changed to set up the single channel analyser to the required channel.

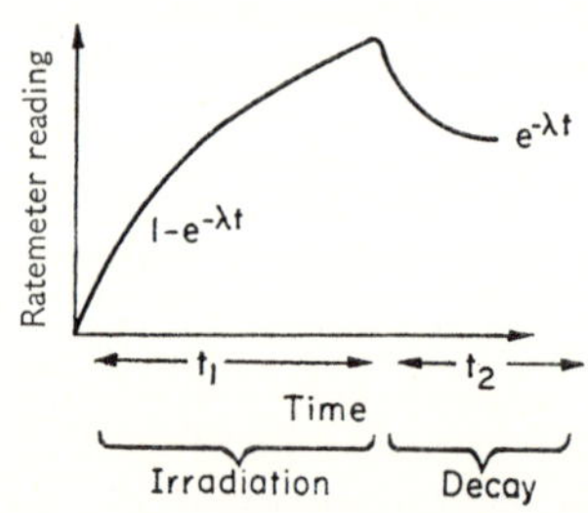

FIG. 40. Neutron monitor ratemeter reading.

The neutron output monitor is simply a thermal neutron detector placed in the shielding. A fission counter is probably one of the best types, as it has a very long plateau, i.e. the count rate is closely constant for large variations in the applied voltage. It is not necessary to find the absolute neutron flux because a standard sample is used for calibration; it is the flux variations which are interesting. The output from the neutron monitor is fed into a ratemeter which is made to have a time constant equal to the time constant of the activity induced. This means that as the irradiation proceeds the ratemeter reading increases in a manner entirely analogous to the increase in activity of the sample (see Fig. 40). On removing the sample the monitor is switched off and the reading drops in the same way that the activity decays. In symbols, the

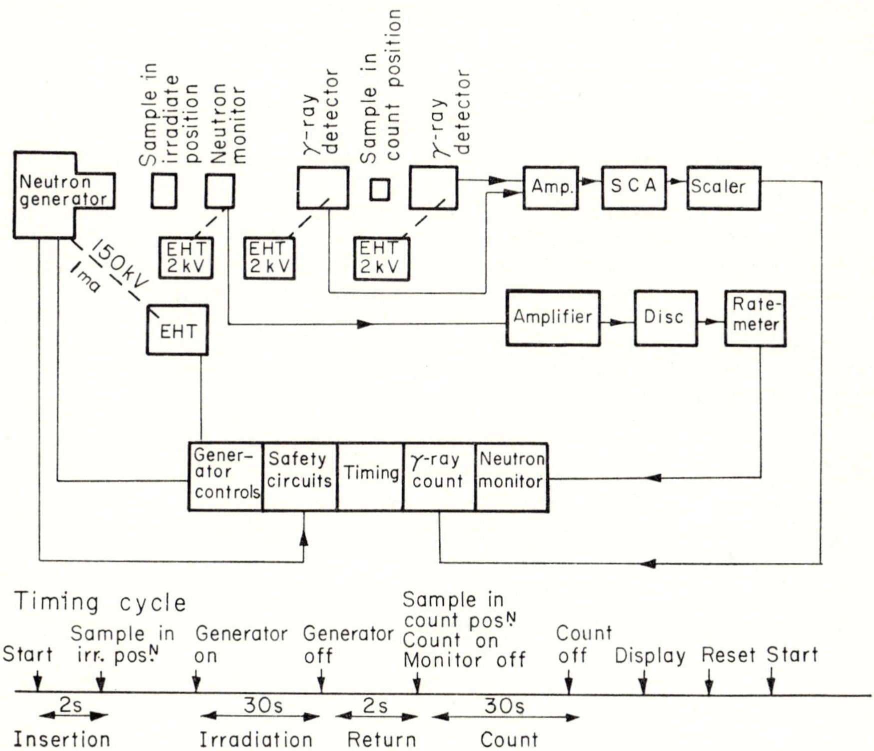

FIG. 41. Schematic diagram for the ANALOX Mark II.

product $\phi(1-e^{-\lambda t_1})(e^{-\lambda t_2})$ is given on the ratemeter. (In systems not employing this device calibration is done by using two samples one of which is a standard. If both are made to spin during the irradiation inhomogeneities are smoothed out.)

About four or five feet of shielding in every direction is required. This may be either water, concrete or wax, each of which has its advantages for particular applications.

Figure 41 shows a block diagram of the "Analox Mk II" steel analysis system,* and also the operation timing cycle. The system is automated to the extent that the operator merely inserts a sample and presses one button. The γ-ray count appears after about one minute and a simple standard calculation gives the oxygen content of the steel. The results appear to be equally as reliable as those from other techniques such as vacuum fusion, chlorination and hydrogen reduction; there are reasons for believing that activation analysis provides the most reliable quick analysis. The time required is in fact less than the time taken to prepare the sample in most cases.

References

1. Morris, D. F. C. (1962). "Neutron Activation Analysis," *Metallurgical Reviews*, **7**, 26.
2. Steele, E. L. and Meinke, W. W. (1962). "Determination of Oxygen by Activation Analysis, etc.," *Anal. Chem.* **34**, 2.
3. Crawford, G. I., Gray, A. L. and Smith, G. D. (1963). "Activation Analysis Applied to Steel Production," *J. Brit. I.R.E.* **26**, 251.

* Manufactured by Plessey-U.K. Ltd., Havant, Hants. The author is grateful to the management of this company for permission to draw on his experience gained with them in preparing this paper.

Chapter 18

ACTIVATION ANALYSIS IN CLINICAL SCIENCE

J. M. A. LENIHAN

Regional Physics Department, 9–13 West Graham Street, Glasgow, C.4, Scotland

The clinical applications of activation analysis are mainly concerned with trace elements. Of the ninety chemical elements that occur in the earth's crust or atmosphere, only twelve are present in substantial amounts in the human body (Table I). Seven other elements (Table II) are

TABLE I

BULK ELEMENTS IN THE BODY

Element	*Amount in 70-kg man (g)**
Oxygen	45,500
Carbon	12,600
Hydrogen	7,000
Nitrogen	2,100
Calcium	1,050
Phosphorus	700
Potassium	245
Sulphur	175
Sodium	105
Chlorine	105
Magnesium	35
Iron†	4

* Best available estimate.

† Some iron in the human body is associated with enzyme systems and behaves as a trace element, but most of the 4-g content is in blood cells and other tissues where it behaves as a bulk element.

known with certainty to have an essential role in the life of man; all of them have been identified in this connection during the last half-century. About twenty other elements occur regularly or occasionally in the tissues of man and other animals as well as in the higher plants. Their biological functions are still uncertain (Table III).

TABLE II

ESSENTIAL TRACE ELEMENTS

Element	*Amount in 70-kg man (g)**
Zinc	1·9
Copper	0·2
Manganese	0·02
Molybdenum	0·015
Iodine	0·03
Cobalt	0·003
Selenium	?

* Best available estimate.

TABLE III

TRACE ELEMENTS

	Date of recognition in tissue of		*Date of recognition of physiological role in*		
	higher plants	*animals or man*	*higher plants*	*animals*	*man*
Al	Early twentieth century	1911	S	L	L
As	Early nineteenth century	1834	L	S	L
B	1895	1912	1926	L	L
Ba	1877	End of nineteenth century	S	S	L
Br	1871	1907	L	L	L
Co	1841	1926	1959	1935	1948
Cd	~1920	~1920	L	S	S
Cr	1911	~1920	L	S	S
Cu	1816	1833	1931	1924	1928
F	Early nineteenth century	1805	L	S	S
I	1811	1895	S	1919	1919
Li	1872		L	L	L
Mn	1913	1913	1912	1931	1930
Mo	1932	1932	1939	1956	1956
Ni	1925	1926	L	S	S
Pb		~1850	L	L	L
Rb	~1880	1882	S	S	S
Se	1880	Early twentieth century	S	1959	1959
Si	~1910	~1910	1926	S	S
Sn		~1900	L	L	L
Sr	nineteenth century	nineteenth century	S	S	L
Ti	~1920	~1920	L	L	L
V	~1879	1913	S	S	S
Zn	1877	1877	1914	1934	1939

L: lack of direct evidence.

S: suggestive evidence, e.g. localization in particular macromolecule, probable role in certain metabolic processes, or improved growth in plants.

For most of the elements in the periodic table, we do not have enough information to assign a nutritional or biochemical role with any confidence. The main reason for this ignorance is the difficulty of detection and estimation of elements present at low levels—often well under 1 p.p.m.

Yet there is good reason to suspect that even very minute concentrations may be significant. Enzyme systems are almost invariably at the mercy of trace element catalysts. Many diseases of great economic significance in plants or animals are known to be influenced by trace elements. Epidemiological studies of human diseases, including dental caries, cancer and coronary thrombosis, suggest that similar mechanisms may possibly be significant in the health of man.

The recognition of an essential role for a particular element depends on four experimental findings:

1. The element must be shown to occur regularly in the body.
2. Association with a particular organ, tissue, enzyme or cell constituent must be established. An element which occurs at measurable concentration throughout the body is probably not a true trace element, but is more likely to be a bulk element.
3. True effects of excess or deficiency of the element must be examined. If no physiological changes result from either a considerable excess or a total deficiency, it is improbable that the element serves any purpose in the body.
4. Finally a rational basis may be found for the presence and behaviour of the element in the body.

The first three stages in this process require analytical techniques of adequate sensitivity. Progress has been slow because so many elements occur in the body at levels near or below the limits imposed by conventional analytical methods.

Activation analysis offers a remarkable sensitivity. For about half of the chemical elements it offers greater sensitivitity than any other method available for the analysis of biological specimens. The gain in sensitivity is often by a factor of 100 or 1000 and for some elements (e.g. arsenic) can be pushed as high as 10^6 without using costly apparatus or techniques after the neutron irradiation. A number of typical applications will illustrate the scope of the method in clinical research.

Enzyme Activity

Trace elements such as Mo, Cu, Se and Zn often form integral parts of enzyme structures. In other instances the enzyme is influenced by a trace element in a fashion not yet fully understood. A clinically important finding arose[12, 13] from the speculation that Zn, which occurs in several enzymes, including alcohol dehydrogenase, might be significant in post-alcoholic hepatic cirrhosis. Investigation showed that zinc metabolism is indeed disturbed in this condition, with reduced concentration in the liver and increased urinary excretion. Limited response to zinc therapy was observed but further study is indicated. It is interesting that the trace elements often found in association with enzyme systems (such as Cu, Mo and Zn) are among those for which activation analysis is very much more sensitive than other techniques.

Epidemiology

The correlation between atmospheric concentrations of various metallic elements and incidence of lung cancer is well established.[11] Arsenic is interesting in this connection because of its occurrence in cigarette smoke. Levels of 50 p.p.m. in cigarette tobacco were observed during the 1950s[1, 5] but the average level in American cigarettes was estimated in 1959[3] at 6 p.p.m. Present (1964) levels are about 1 p.p.m.

Even so, the smoker probably deposits appreciably increased amounts of arsenic in his lungs. It is known (from experiments with cigarettes labelled with ^{74}As) that about 3% of the arsenic in a cigarette is eventually deposited in the lungs.[5] The smoking of one cigarette will deposit about 0·03 μg of arsenic in the lungs. In a typical British environment the atmospheric arsenic level is about 50 $\mu g/1000\ m^3$.

A simple calculation suggests that the total amount of arsenic inhaled in a day is about 1 μg and that the total amount deposited in the lungs (assuming that inhaled air and cigarette smoke behave similarly in this regard) is about 0·03 μg. A moderately heavy smoker may therefore increase his lung burden of arsenic quite substantially, even at today's greatly reduced concentration of this element in tobacco. Further work is in progress, including the analysis of large numbers of tissue samples.

Dental caries offers another fascinating epidemiological problem. A protective role has been suggested for a number of trace elements, including Mo, Sr and V.[10, 6, 7] The widely different incidence of the disease in the neighbouring (and very similar) New Zealand towns of Hastings and Napier may be linked with the difference in trace element content of locally-grown vegetables, or as a result of soil changes following volcanic action. In Napier, where the disease is less prevalent, the diet appears to be richer in Al, Mo and Ti.

Other possibilities—still speculative, but inviting the attention of activation analysts—include the role of trace elements in coronary heart disease, now being studied in Glasgow.

Toxicology

Arsenic is still significant as a poison—administered with homicidal purpose or ingested accidentally by industrial workers (e.g. in the manufacture of sheep-dip or insecticides). The incidence of lung cancer and skin cancer is high in factory workers handling sodium arsenite.[4] Medicinally administered arsenic also appears hazardous.[8, 9]

For other metallic poisons, including Hg, Mn and Sb, the superior sensitivity of activation analysis opens the way for further investigations which are now in progress; Hg and Sb are being studied in Glasgow.

Future Developments

The main need at present is for systematic information on the normal levels of many elements in the different tissues of the body. Measurements of this kind are tedious but the work is proceeding. Additional prospects arise from the use of particles other than neutrons. Fleckenstein and Janke[2] have described an elegant method for the study of oxygen in intra-cellular metabolism. They use the stable isotope ^{18}O as a tracer, estimating it by activation with protons in the reaction $^{18}O(p,n)^{18}F$ to give the fluorine isotope of half-life 1·9 hours. Some of the stable isotopes of other elements (such as Ca and Sr) which normally occur at very low levels in nature, may be concentrated by electromagnetic separation and used as tracers in biological systems. Their eventual assay, which would be difficult by traditional methods, can be achieved by activation analysis in many instances.

In general, the methods of activation analysis have much to offer the clinical research worker. Their wider application in medicine seems likely to produce valuable results of a practical kind.

References

1. Bailey, E. J., Kennaway, E. L. and Urquhart, M. E. (1957). "Arsenic Content of Cigarettes," *Brit. J. Cancer*, **11**, 49–53.
2. Fleckenstein, A. and Janke, J. (1964). "L'Application biologique de l'analyse d'activation de l'oxygène 18," In *L'Analyse par radioactivation et ses applications aux sciences biologiques*. Presses Universitaires de France, 267–285.
3. Guthrie, F. E., McCants, C. B. and Small, H. G. (1959). "Arsenic Content of Commercial Tobacco, 1917–1958," *Tobacco Science*, **3**, 62–64.
4. Hill, A. B. and Faning, E. L. (1948). "Studies in the Incidence of Cancer in a Factory Handling Inorganic Compounds of Arsenic," *Brit. J. Industr. Med.* **5**, 1–6.
5. Holland, R. H., Wilson, R. H., Acevedo, A. R., McCall, M. S., Clark, D. A. and Lanz, H. C. (1958). "A Study of Arsenic in Regular-Sized Unfiltered and Filtered Cigarettes," *Cancer*, **11**, 1115–1118.

6. Kruger, B. J. (1959). "The Effect of 'Trace Elements' on Experimental Dental Caries in the Albino Rat," *University of Queensland Papers*, **1**, 3–28.
7. Ludwig, T. G., Healy, W. B. and Losee, F. L. (1960). "An Association between Dental Caries and Certain Soil Conditions in New Zealand," *Nature*, **186**, 695–696.
8. Neubauer, O. (1947). "Arsenical Cancer: a Review," *Brit. J. Cancer*, **1**, 192–251.
9. Robson, A. O. and Jelliffe, A. M. (1963). "Medical Arsenic Poisoning and Lung Cancer," *Brit. Med. J.* **2**, 207–209.
10. Rygh, O. (1949). "Importance of Trace Elements in Nutrition," *Research*, **2**, 340–341.
11. Stocks, P. (1960). "On the Relations between Atmospheric Pollution in Urban and Rural Localities and Mortality from Cancer, Bronchitis and Pneumonia with Particular Reference to 3:4 Benzopyrene, Beryllium, Molybdenum, Vanadium and Arsenic," *Brit. J. Cancer*, **14**, 397–418.
12. Vallee, B. L., Wacker, W. E. C., Bartholomay, A. F. and Hoch, F. L. (1956). "Zinc Metabolism in Hepatic Dysfunction," *New Engl. J. Med.* **255**, 403–408.
13. Vallee, B. L., Wacker, W. E. C., Bartholomay, A. F. and Hoch, F. L. (1957). "Zinc Metabolism in Hepatic Dysfunction," *New Engl. J. Med.* **257**, 1055–1065.

Chapter 19

FORENSIC APPLICATIONS OF ACTIVATION ANALYSIS

V. P. GUINN

General Atomic Division, General Dynamic Corpn.,
P.O. Box 608, San Diego, California, U.S.A.

The use of activation analysis in forensic work can have three distinct advantages over routine chemical methods: firstly, the sensitivity attainable with high neutron fluxes permits analysis of very small samples; secondly, the analysis of larger samples can be carried out for lower quantities of trace elements; thirdly, the method is often non-destructive, thus allowing the sample specimen to be presented as evidence in courts of law.

Among the materials so far analysed in forensic work are: gunshot residues, coins, hair, glass, greases, wood, soils, drugs, plastics, pesticidal residues, cigarettes, tyres, oil, rubber, paper, plastics, tape, putty, inks, whisky, and paints. A brief account of the case work involved for several of these is given.

Gunshot Residues

Every time a pistol is fired, a small amount of detonating material is blown back past the hammer and onto the back of the firer's hand between the base of the thumb and forefinger. This residue can be removed by applying molten paraffin wax to the hand and removing it after solidification. Treatment of the cast with diphenylamine and concentrated sulphuric acid gives a blue colouration when nitro-compounds are present. Unfortunately, this chemical test is not reliable. The activation method in this case involves the production of radioactive isotopes of barium and antimony, elements which are present as constituents of cartridge primers.

The normal range of barium and antimony levels on a "clean" hand, and the levels after a typical single firing of either 0·22 calibre or 0·45 calibre revolver, are tabulated below

	Ba	Sb
Normal level	0·13 ± 0·11 μg	0·015 ± 0·012 μg
0·22 calibre	0·40 ± 0·18 μg	0·07 ± 0·01 μg
0·45 calibre	5·6 ± 3·6 μg	0·96 ± 0·82 μg

The contamination is seen to be considerably greater for the larger calibre revolver. The wide tolerances obtained result from the many factors governing the amounts of these elements deposited on the hand. These include wind direction, type of weapon, amount of moisture on the hand, and whether the gun was clean or dirty. Washing does remove the residue but this can be an advantage, since if the suspect is arrested fairly soon after the crime, any earlier firings are of no significance.

It might be very convenient if every manufacturer were to label his cartridges with tracer elements in the gunpowder, using a combination of tracers to denote origin and calibre of the cartridge. The obstacle to this proposal is the fact that many manufacturers obtain their materials from the same suppliers and may possibly change their supplier from time to time. However, there is a good possibility that manufacturers will agree to using a single tracer element—one that is more uncommon than Ba and Sb, and that is even more readily determined by NAA.

The same technique may be applied to the problem of narcotics, which could be labelled with low level trace elements. Routine analysis of all drugs confiscated during arrests, or obtained from known sources, would help to build up a picture of the trafficking routes. Overtures are being made at present on an international level and it is hoped that an operating system can be achieved.

One very interesting case in which the analysis by activation of gunshot residues has been used is that of the assassination of President John F. Kennedy. The suspect, Lee Harvey Oswald, who was himself killed in the same manner, was arrested after having shot and killed a police officer. Thus, since the assassination of the President was committed with a rifle, the conclusive factor would be the presence of residues not only on the gun hand but on the face and possibly both hands, since a rifle is fired with the face in the vicinity of the breech.

To this end an Italian rifle, identical in type to the one allegedly used by Oswald, was purchased from the same store—in a study carried out jointly by the Los Angeles Police Department, and the author's laboratory. A series of tests was made to determine the levels of residues on the cheeks and hands of the person firing it for various numbers of shots. Paraffin casts were made, some tested then by the diphenylamine procedure, others by NAA. The diphenylamine results were useless, but the NAA method clearly detected both Ba and Sb, on both cheeks and both hands of the firer each time. The chemically-tested casts were then also examined by NAA; Sb was still found, but not Ba. Subsequently, the F.B.I. examined the Oswald casts, which had been tested by diphenylamine in Dallas, by this NAA procedure. (Note added later: as described in the Warren Commission Report, the NAA measurements showed the presence of Sb and Ba on both hand casts, and the one cheek cast of Oswald, but the results were inconclusive, as they were found on the outside as well as inside surface of each cast, presumably indicating careless handling of the casts at some earlier time.)

Another example of the application of activation analysis to forensic work began when the U.S. Treasury Department was approached by a numismatist who claimed to have in his possession a number of 1943 American copper pennies. Copper was used in minting pennies in 1942, but was exchanged for nickel-steel in 1943. Activation analysis carried out on the 1943 copper penny, which was otherwise perfect, showed that the trace elements used in it were present at levels identical to those in the 1942 pennies. It was therefore concluded that the penny was genuine. It has been speculated that an employee of the U.S. mint had used mint copper to produce a few collector's pieces early in 1943. There are now known to be five such coins in existence, valued at around $50,000 each.

Other cases include that concerning the conviction of a man suspected of despatching a bomb via the U.S. Mail. Adhesive tape and automobile putty used to parcel the bomb were matched, by the U.S. Internal Revenue Service activation analysis group, with similar materials found in the suspect's garage.

The first test case in which evidence for the prosecution rested almost solely on activation analysis findings was carried out by the author's group, and was reported recently (Time magazine of August 7th, 1964). The charge of attempted burglary was against a private investigator whose licence to operate would be revoked in the event of a judgement against him. The suspect was arrested, essentially at the scene of the crime, by police officers who answered an alarm call. A car jack-handle was found near the scene and the suspect's car was parked nearby, but there was no otherwise conclusive evidence that the suspect was in fact guilty.

On inspection, the jack-handle was found to bear traces of paint at two points, blue paint on the body of the handle matched qualitatively with that on the boot of the suspect's car, and brown paint on the wedge end of the tool matched in appearance that on the door which had been forced. These paint traces were too minute (10 μg, 50 μg) to be analysed by any other available technique than activation analysis, which showed that the detected element levels—seven elements in the brown paint and five in the blue—matched quantitatively to a high degree of accuracy. Presentation of this evidence, with the explanation that the probability that they had the same origin (same brand of paint) was around 99·99%, secured a verdict in favour of the prosecution.

Chapter 20

ACTIVATION ANALYSIS IN THE PETROLEUM AND CHEMICAL INDUSTRY

V. P. GUINN

General Atomic Division, General Dynamics Corpn., P.O. Box 608, San Diego, California, U.S.A.

The oil and petrochemical industry deals basically with one type of compound—hydrocarbons, which are particularly suitable for activation analysis experiments, and simple derivatives of hydrocarbons. Since the hydrocarbon consists only of carbon and hydrogen, elements which are not sensitive to either fast-neutron or thermal-neutron activation, one has essentially a transparent matrix. This means that impurities, such as halogens and trace metals, stand out strongly, and can be easily estimated. Owing to this ease of estimation, it is often quite possible to obtain good sensitivity with simple, and relatively inexpensive, neutron generators, which most companies can afford. On the other hand, if a nuclear reactor had been required for the experiments in question, it is unlikely that the petrochemical industry would be to the fore in this field.

Crude oil contains essentially free hydrocarbons; as impurities there are many organic compounds containing oxygen, sulfur, and nitrogen, along with a large number of trace metals such as nickel, iron, manganese and sodium, usually at levels in the range of 1–100 p.p.m. The crude oil is often particularly rich in vanadium, which can be present at levels up to even 1000 p.p.m. (It is interesting to note that many of the metals are present in porphyrin-type structures.) For a satisfactory fuel, a product containing essentially nothing but hydrocarbons is usually desired. It is essential, therefore, that the oxygen, nitrogen and trace metals be removed. The standard method of eliminating most of the metals is by vacuum distillation, distilling over the hydrocarbons, and leaving behind a residue high in metals. This is straightforward in principle, but in practice certain non-volatile materials do distil over or are carried over by entrainment. This would not be too important but for the fact that the next step in the processing of the crude oil is usually that of catalytic cracking. This catalyst is fairly expensive, and it is essential that it remain active and selective as long as possible. A sizeable fraction of the non-volatile material which does entrain over contains nickel and vanadium compounds, and both of these elements are catalytic poisons. Owing to the expense of the catalyst, it is essential that these metals do not come over in any appreciable quantity, and a good way to ascertain this is to use activation analysis. The method for determining vanadium is fast, since the isotope measured is vanadium-52 with a half-life of only 3·8 min. Also, it turns out that vanadium is often the only metal whose level need be measured, since the levels of the others are in many cases approximately proportional to it.

Another metal which quickly poisons many catalysts is arsenic. This is not found in the crude oil itself, but is often added to the oil wells as a corrosion inhibitor for the steel pipes of the well. High-flux (reactor) activation analysis checks the arsenic content in the feed from the purification units, which remove this element from the crude oil. The purified feed should be less than 0·001 p.p.m. in arsenic if it is to be used in a platforming unit—otherwise it will soon poison the very expensive platinum catalyst.

Activation analysis is also used in a number of related industries. In particular one should mention the plastics industry, whose raw materials are often ethylene and propylene, obtained from the oil industry. Plastics are often made from such olefins by catalytic polymerization, in many cases using Ziegler's catalyst (a complex chloride of titanium and aluminium). To start the free radical polymerization process, a molecule of catalyst forms a free radical monomer in conjunction with the organic unit. This means that the final large polymer molecule will have at least one molecule of catalyst in it; the concentration of the metal in the polymer will probably be in the p.p.m. range, and hence suitable for estimation by activation analysis. If the polymer is not properly washed, of course, the levels will be much higher, and deleterious to the properties of the polymer. The sensitivities of the appropriate reactions for chlorine and aluminium are such that they can be readily detected whereas titanium can only be detected at higher levels—using a small neutron generator. It should be noted that these elements are a source of impurity and error when activation analysis samples are irradiated in polythene tubes unless a container blank is determined and subtracted, or the samples are transferred before counting.

An example of the use of activation analysis in the research laboratory is in the determination of the empirical formulae of new compounds. Conventionally, when a new organic compound is prepared, the carbon and hydrogen contents are estimated by combustion methods. Any oxygen present is usually computed merely by difference, and the result is thus necessarily of limited accuracy, since it involves the sum of the errors of the previous determinations.

Activation analysis can be used to determine the oxygen content, directly and accurately, in samples as small as 10 mg by the fast neutron reaction:

$$^{16}O(n, p)\,^{16}N$$

It should be noted that the technique is non-destructive and very rapid (since the half-life of ^{16}N is only 7·35 sec). In an analogous manner, nitrogen can be estimated using the reaction:

$$^{14}N(n, 2n)\,^{13}N$$

An interesting difficulty appears with this reaction. In a recent investigation, a calibration curve, measuring counts per minute against known amounts of nitrogen was plotted, Fig. 42. The expected curve would be

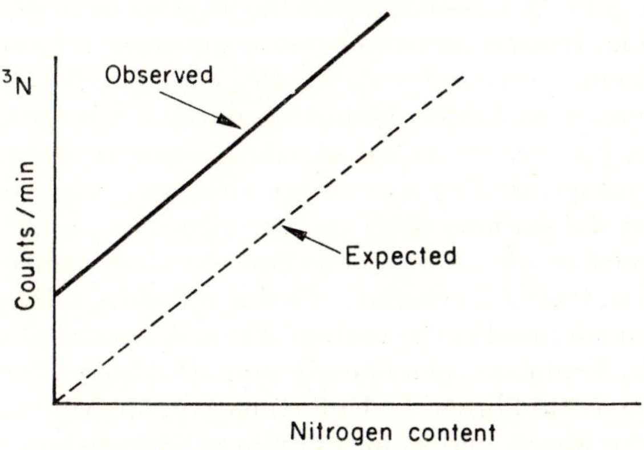

FIG. 42. Nitrogen-13 calibration curve.

expected to go through the origin; unhappily the curve obtained did not, and it was assumed that the matrix compound used (a supposedly very pure hydrocarbon) was impure, and contained some contaminant with a higher nitrogen content. After repeated attempts to purify the compound, the same curve was still obtained, and it was then realized that some other effect was causing the abnormality. What was happening was that a proton recoil reaction was taking place, and this also formed the nitrogen-13 nuclide through the reaction,

$$^{13}C(p, n)\,^{13}N$$

This gives the appearance of about 400 p.p.m. nitrogen in a typical pure hydrocarbon.

Pesticide residues in foodstuffs also provide an interesting field for activation analysis. These compounds are essential if we are to obtain good crop yields. Unfortunately many pesticides are also poisonous to man; since they can be incorporated into the crop, it is necessary to establish safety levels. For many pesticides, activation analysis is not an immediate help, since one is attempting to estimate part of an organic molecule in a complex organic matrix. Generally, in such cases, it is first necessary to carry out some kind of extraction procedure, the extract then being activated. If this is done, one can detect even 0·001 p.p.m. of DDT, using an (n,γ) reaction on the chlorine atoms in the molecule (in the reactor). Bromine compound residues are determined more readily—directly and instrumentally. One can thus measure bromine down to the 1 p.p.m. range by counting the 36-hour Br^{82}. Over application of bromine-containing pesticides can raise the level of this element in foodstuffs to levels of the order of 100 p.p.m.; since many people are allergic to more than 10–50 p.p.m., it is essential that the level be reduced, preferably to about 10 p.p.m. In some pineapple-producing areas, it has been common practice to fumigate the soil with bromine compounds such as dibromochloropropane for protection against nematodes. The bromine level has been building up over the years, so that the pineapple product now in some areas contains 10–20 p.p.m. of bromine, and may eventually reach the danger level if suitable steps are not taken.

Wheat stored in silos needs to be fumigated at regular intervals to combat insect attack. The reagent used is normally methyl bromide, a gas, which is introduced at the bottom of the silo, and allowed to work its way upwards. Vents are then opened, and air is blown through the silo to remove the methyl bromide. Unfortunately, some of the gas actually reacts with the wheat, and so forms organic bromine compounds. With regular fumigation, it is quite possible to build up the bromine content to over the tolerance level of 50 p.p.m. Again, levels of bromine > 1 p.p.m. are readily determined by instrumental NAA.

Seed wheat is treated with mercury compounds, to the extent of about 10 p.p.m. Hg. These compounds act as fungicides when the seeds are first placed in the ground. If the farmer has excess seed wheat, he may decide to add it to his crop at the end of summer, rather than throw it away, although this is strictly forbidden by American law, because of the danger of mercury poisoning. Activation analysis with a reactor can measure mercury down to even 0·001 p.p.m., and so tell whether this forbidden practice has been carried out.

Chapter 21

ACTIVATION ANALYSIS IN DENTAL MEDICINE

G. S. NIXON

University of Glasgow Dental Hospital and School,
211 Renfrew Street, Glasgow, C.3, Scotland

In dental medicine, activation analysis is proving a most useful research tool, and one which is contributing towards an understanding of many controversial problems, particularly that of dental caries.

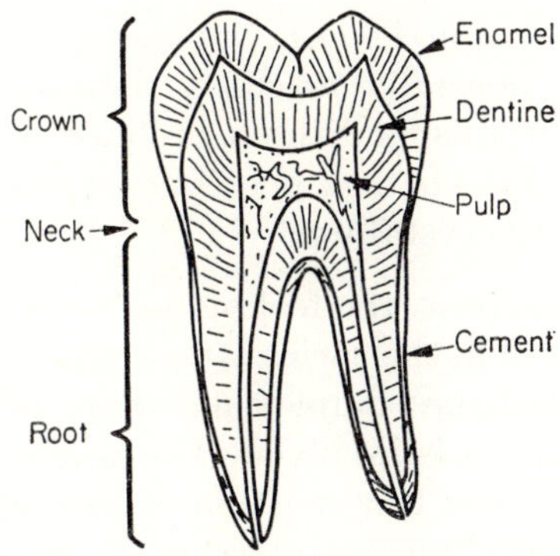

FIG. 43. Longitudinal section through tooth showing the dental tissues.

The structure and composition of the hard tissues of the teeth are shown in Fig. 43.

Tissue	*Composition*	
	Inorganic	*Organic*
Enamel	98%	2%
Dentine	75%	25%
Cementum	68%	32%

Dental caries is, without doubt, a disease of civilization. It is a disease which is increasing each year despite increased dental care. Many factors, such as racial and dietary factors, are involved in the mechanism of dental caries. There is also undoubtedly a close relationship between the consumption of refined food and the incidence of caries. In races where little refined food is eaten, there is little development of caries. It is also a well-known fact that the increase in caries has followed closely an increased consumption of sugars. A minority have suggested that refining of food-stuffs removes essential nutritional factors, including trace elements, from the diet. This hypothesis has been strengthened by the finding that trace amounts of fluorine in the water have an inhibiting effect on dental caries.

The mechanism of caries is still unresolved, and the three main theories are:

(i) Acidogenic Theory, which postulates that oral bacteria readily break down sugars to acid end-products, which then attack the inorganic constituents of enamel.
(ii) Proteolytic Theory, whereby the attack is made primarily on the organic constituents of enamel by bacteria which produce enzymes capable of digesting the organic matrix.
(iii) Proteolytic-Chelation Theory, which postulates that the organic matter in the normally calcified tissues can be attacked by certain micro-organisms without preliminary decalcification, and that products of the attack possess complexing properties which affect the apatite crystallites.

Unfortunately, in the past, the feeling has been that one mechanism, and one only, is involved in the caries attack. Recent research, however, has served to emphasize the complex nature of caries, suggesting that more than one mechanism is involved. The complexity of the problem becomes very evident when the question of individual susceptibility to dental caries is examined. Frequently patients' mouths are examined and found to be completely caries-free, where no attention has been paid to oral hygiene.

At first sight, it would appear to be an obvious approach to the caries problem to investigate the composition of the enamel in relation to caries resistance as enamel is the first tissue to be attacked by the carious process. This approach has, unfortunately, several disadvantages, such as the basis for deciding which teeth are caries-free or caries-prone. Until caries has begun and produced chemical alteration, there is no means of knowing whether a particular tooth is caries-prone. Likewise, if one takes a caries-free tooth, there is no guarantee that this tooth would have remained so had it remained in the mouth for a longer time. It must be emphasized that the composition of the tooth is one of the important factors under investigation but it is not the only factor involved in caries susceptibility. The early literature contains suggestions of differences in composition of the primary elements in the inorganic fraction between sound and carious teeth, but the results are of limited value as whole teeth or roots and crowns were usually employed rather than samples of enamel or dentine.

There are many indications that the trace elements do play an important part in caries-resistance. Sognnaes[1] investigated the nature of the substances present in natural diets which reduced the caries susceptibility of animals, when fed during the period of formation of the teeth. They suggested that some mineral substances present in the diet, which they called the "natural diet", were possibly trace elements which increased the resistance of the enamel. In addition to fluorine, which was mentioned briefly, other trace elements, vanadium, molybdenum and manganese, are believed to increase resistance to caries, while selenium has been said to be associated with an increase in caries. It is interesting to note that the ash which Sognnaes gave to his rats contained fluorine with vanadium and molybdenum. The concentration of trace elements such as these has not been adequately studied.

There is also a close relationship between the presence of trace elements in soil and water, and the incidence of dental caries. In 1953, it was observed in Hungary by Adler[2] that children in some districts, in which small amounts of molybdenum were present in the water, had less caries than children from other districts where the water supply did not contain appreciable amounts of molybdenum. A similar situation has been observed in New Zealand. In 1931, an earthquake near the town of Napier caused the raising of the soil in this area. Subsequent dental examination, in 1954–55, of children from this district showed that they had less caries than children from the nearby town of Hastings. The water supplies of the two towns were similar in composition, but examination of vegetables grown in Napier showed that there was a significant difference in the trace element content of the edible part of the vegetables and that this difference lay in the molybdenum content.[3] Experimentally, it has been shown that there was a reduction in dental caries in rats fed on experimental diets to which the ash of vegetables grown in Napier had been added.

It has been suggested that the trace elements, such as vanadium, replace phosphorus in the hydroxy-apatite crystals and, when incorporated in enamel and dentine, increase the hardness of the hydroxy-apatite, as well as the cohesion between organic and inorganic matter. It has also been suggested that trace elements (such as vanadium and fluorine) when applied to enamel surfaces form compounds with the hydroxy-apatite crystals which are effective in reducing the acid solubility of enamel.

The earlier studies which established trace element levels in teeth, used conventional methods of analysis. One of the main difficulties with these methods was in sampling, owing to the large amount of sample necessary and also the necessity of separating the hard tissues (enamel and dentine). To obtain sufficient samples, it was necessary to use pooled samples from as many as twelve teeth, often obtained from different individuals. The advantage of the activation analysis method is that small samples can be used, which may be obtained from individual teeth. In addition, different layers of enamel and dentine can be obtained from individual teeth. In preparing a sample of tooth for activation analysis, the separation of enamel and dentine is difficult and has to be carefully carried out. After cross-sectioning the crown of each tooth, the enamel and dentine thicknesses are measured. The enamel is sectioned into either outer and inner, or outer, middle, and inner layers, using a fine carborundum disc (0·005 in. or 0·01 in.) on a micrometer cutting gauge. Thus, from each tooth, a number of samples of each layer can be obtained. To avoid contamination, a fresh carborundum disc is used for each cut.

Using activation analysis methods, arsenic, copper and mercury have been estimated in enamel samples. These results have shown that differences in concentration in the trace elements are found between different layers of enamel. They have also shown that individual variations for these elements do occur.

One interesting application was to determine the mercury content of the enamel of sound unerupted and erupted teeth and to determine if migration of mercury ions took place from silver amalgam fillings present in other teeth. Enamel samples were obtained from permanent human teeth which were free from enamel defects. None of the teeth in this group was in direct contact with teeth containing amalgam fillings. In this way, direct contamination of the enamel with mercury from amalgam was avoided as far as possible. Unerupted teeth which, on clinical examination showed no communication with the oral cavity, were also used. A further series of enamel samples were taken from sound teeth in direct contact with teeth containing silver amalgam fillings. The mercury content found was:

Mercury in Enamel, p.p.m.

Type	*Inner Layer*	*Outer Layer*
Unerupted teeth	0·10	0·10
Erupted teeth	2·34	2·79
Teeth in immediate contact with silver amalgam fillings	625·00	977·00

In the early stages of investigation, elements with long half-lives were chosen because of the difficulty of irradiation, but with the establishment of the Scottish Universities Research Reactor, work has proceeded with shorter-lived isotopes. One of the first to be investigated was vanadium. This trace element had already been studied by Söremark[4] in biological material. Vanadium appears to be in a concentration of less than 0·05 p.p.m. in the enamel of teeth in the West of Scotland.

Methods have now been established for the trace elements, molybdenum, zinc, antimony and manganese, and their concentrations in teeth are now being examined.

Activation analysis is also being used to determine the relationship between the formation and composition of dental calculus (tartar) and saliva.

References

1. Sognnaes, R. F. (1948). "Caries—Conducive effect of a purified diet when fed to rodents during development," *J. Amer. Dent. Ass.* **37**, 676–692.
2. Adler, P. (1957). "Experiments with albino rats upon the caries effect of water-borne molybdenum," *Odont. Revy.* **8**, 202–207.
3. Ludwig, T. G., Healy, W. B. and Losee, F. L. (1960). "An association between dental caries and certain soil conditions in New Zealand," *Nature*, **186**, 695–696.
4. Söremark, R. (1963). "Studies on the cencentration of vanadium in some biological specimens," Proc. 3rd Int. Conf. Biol.

Further Reading

1. Brudevold, F. (1962). "Dental tissue factors in dental caries," *Internat. D. J.* **12**, 496–507.
2. Ericsson, Y. (1962). "Recent advances in dental caries research. Biochemistry, salivary and food factors in dental caries development," *Internat. D. J.* **12**, 476–495.
3. Jenkins, G. N. (1961). "The biochemistry of enamel in relation to caries resistance," *Arch. Oral Biol.*, Supplement Vol. 6, 305–314.
4. Kruger, B. J. (1959). "The effect of 'trace elements' on experimental dental caries in the albino rat," *Univ. Queensland Papers, Dept. Dent.* **1**, 28.
5. Mühler, J. C. (1957). "The effect of vanadium pentoxide, fluorides and tin compounds in the dental caries experience in rats," *J. Dent. Res.* **36**, 787–794.
6. Shaw, J. H. and Griffiths, D. (1961). "Developmental and post-developmental influences in incidence of experimental dental caries resulting from dietary supplementation by various elements," *Arch. Oral Biol.* **5**, 301–322.
7. Söremark, R. and Samsahl, K. (1961). "Gamma-ray spectrometric analysis of elements in normal human dentine," *J. D. Res.*, **41**, 603–606.

8. Söremark, R. and Samsahl, K. (1961). "Gamma-ray spectrometric analysis of elements in normal human enamel," *Arch. Oral Biol.*, Supplement Vol. 6, 275–283.

9. Söremark, R. and Samsahl, K. (1962). "Analysis of Inorganic Constituents in Dental Calculus by means of neutron activation and gamma-ray spectrometry," *J. Dent. Res.* **41**, 596–602.

10. Tank, G. and Storvick, C. A. (1960). "Effect of naturally occurring selenium and vanadium on dental caries," *J. Dent. Res.* **39**, 473–488.

Chapter 22

HANDLING OF BIOLOGICAL MATERIAL

HAMILTON SMITH

Department of Forensic Medicine, The University, Glasgow, W.2, Scotland

Activation analysis is a very simple technique which allows small samples to be assayed for trace elements. Because we are attempting to estimate very small quantities, contamination before irradiation is a serious problem and one which can only be kept in bounds by strict attention to clean handling, or preferably no handling at all. The latter state can only be achieved in special circumstances, e.g. the irradiation of peas in pods. Normally we must take the greatest amount of care possible.

We can distinguish between two types of biological material, wet and dry. We shall consider the former first, since most biological tissue is wet, and the trace elements in it are fairly evenly distributed. If we wish to analyse a piece of human tissue, it must be removed from the body, usually not by the analyst, and here the first source of contamination is found. This contamination comes from a surprising variety of sources; (i) the surface of the body (dust, dirt, industrial contamination, cosmetics); (ii) the gloves of the pathologist (disinfectants, talc, dust); (iii) the instruments used (metallic dust from the instrument, contamination from previous organs dissected); and (iv) the dissecting table (dust, disinfectant). Having been removed from the body, the tissue is then placed in a clean polythene container without the addition of any preserving chemicals, which would be a rich source of contamination.

If the sample is large enough, it is good practice to use a piece from the centre of the organ taking care to avoid further spread of contamination. The piece can be removed by a clean knife and placed in a polythene container immediately. It is almost certain that the cut tissue will be contaminated by the material of the knife, but if a chemical separation is proposed after irradiation, this will not matter if a knife is chosen which is not made of the element in which interest lies. If a general analysis is proposed, plastic instruments should be used, but these are not satisfactory since they possess poor cutting edges. Where the piece of tissue is too small to allow further cutting, there is not much that can be done, other than analysis by chemical separation for a species not likely to be present as a major contaminant.

Fortunately, biological material has a wide range of normal values for most trace elements—up to a hundred times the minimum value, so significant variations are not noticed with minor contaminants. In closely controlled research, where small variations are important, the necessary steps to control contamination can be taken.

It is always more convenient to work with dry tissue. This material can be obtained in a variety of ways, but principally by vacuum drying techniques using refrigeration or drying agents. With the latter, of course, it is necessary to take great care, and use only those which give the minimum contamination or which do not contain the species being sought. All tissues can be treated in this way, even blood and urine. When dry, the tissues may be stored for many months in thin polythene envelopes without fear of decay.

Dry tissue also includes the outer coverings of the body—skin, hair, and nail. This tissue is open to continuous contamination, and its trace element content is dependent on exposure. An example of this is that hair should give similar analytical results whether it is body or head hair as both receive trace elements from the blood stream initially; head hair invariably shows higher levels of contamination since it is exposed to the atmosphere. A more practical example is that of the mercury content of the toe and finger nails of dentists' assistants. The content of toe was approximately normal since mercury gets there by absorption only. In the finger nails, the metal content is exceptionally high from working with mercury amalgams. Trace elements in dry tissue come from two sources: (i) internal deposition from the blood stream—this is a reflection of the diet, and possibly respired air; and (ii) external contamination from the atmosphere (industry dependent), cosmetics and detergents, and probably the soil.

Attempts have been made to remove external contamination of tissue but this is often, if not always, impossible without partial destruction of the tissue and exposure to further contamination from the cleaning materials. The only cleaning that should be attempted is removal of obvious particulate contamination by scraping with a perspex knife and rinsing with de-ionised water.

The technique of tissue ashing can be used to concentrate trace elements, but it is of limited application, as a large number of elements are lost, or partially lost, by volatilization, and the sample is open to contamination from the materials of the crucible and the muffle furnace.

If it is used, trace element experiments should be carried out to prove its validity.

A brief mention of laboratory handling is appropriate. Where possible, stainless steel knives and forceps should be used, but it will sometimes be necessary to use the less satisfactory plastic instruments. The samples should be received and processed in a room which is substantially dust free, and certainly free from chemical contamination. Sets of apparatus are kept in this room for the purpose of tissue handling, and are not used otherwise. The usual materials used in the packing of samples for irradiation are aluminium, silica and polythene. Aluminium is usually cleaned with benzene (to remove grease) and then with very dilute nitric acid followed by rinsing with water. Samples may be placed in thin aluminium tubes with slip-on lids, or they may be wrapped in aluminium foil. As these containers are not sealed, they are unsuitable for irradiation in which there may be escape of volatile substances. Aluminium packing materials are suitable for long irradiations. Silica is usually cleaned with a mixture of sulphuric and nitric acids (3:5) followed by rinsing with distilled water. Samples are placed in silica ampoules of convenient size, which are sealed with an oxygen/gas flame. Ampoules of this type are suitable for all irradiations in a reactor and for all liquid samples. Polythene is cleaned by the same method as silica or it may just be washed with distilled water: this appears to be quite satisfactory. Samples may be placed in small sealed envelopes of thin polythene, if the irradiation temperature is less than 60° C; above this temperature, the envelopes tend to stick together, giving an inseparable mass. Alternatively, they may be placed in polythene tubes of convenient thickness and length. These may be sealed with a polythene plug, welded to the sides by heating with a small flame.

The amount of material irradiated depends on the sensitivity of the activation and the expected trace element content. We have used samples of tissue varying from 3 μg to 50 mg; liquid samples usually weigh about 0·1 g. The number of samples irradiated at one time is usually regulated so that separations can be completed in about four half-lives. At thermal neutron fluxes of about 10^{12} neutrons/cm^2/sec, there is generally a large enough volume available to take about fifty 50-mg samples. In high-flux reactors (say 10^{14} neutrons/cm^2/sec), the space available is limited; a typical irradiation unit is a silica ampoule 25 mm long with an inside diameter of 3 mm. This limits the number and size of the samples and makes analysis of large numbers difficult.

The time for which a sample can be irradiated also varies. Dry tissue samples can stand irradiation by 10^{12} neutrons/cm^2/sec for a week, without much damage, when the reactor temperature is about 100° C.

Hair is an exception; it becomes brittle if irradiated for more than a day, and disintegrates almost completely if irradiated for a week. At high flux densities, suitable for analysis of short sections of hair, damage can be so great that the hair is "burned" onto the container walls. Liquid samples must always be packed in silica ampoules, and these should be tested under the temperature conditions likely to be experienced in the reactor. Urine samples, packed in such a way, suffer no damage after one day in the reactor. Serum sets to a jelly-like consistency after about twenty minutes, and is then difficult to remove from the ampoule. Usually it is necessary to break the ampoule and dissolve out the contents. This, however, means that the ampoule must be well cleaned to remove external contamination of the silica.

Finally, since we are concerned with biological material, it seems appropriate to say a few words about the problem of digestion. By digestion we mean the destruction of the organic matrix, and the dissolution of the trace elements. When concentrated sulphuric acid is used, a black homogeneous solution is obtained after digestion, but this clears to a yellow solution after the addition of sodium nitrate. Alternatively, concentrated nitric acid can be used. Unfortunately large quantities are required, but fairly low temperatures only are required for digestion. Probably it is best to use a mixture of both acids in suitable proportions. Other chemical reagents have been used for digestion, but many of these are of limited application.

Chapter 23

TRACE ELEMENTS IN PLANT AND ANIMAL TISSUES

H. J. M. BOWEN *

*United Kingdom Atomic Energy Authority,
Wantage Research Laboratories, Wantage, Berks., England*

Activation analysis has a number of inherent advantages and disadvantages in the study of biological material. The three most important advantages are: (*a*) sensitivity; (*b*) freedom from contamination; and (*c*) the possibility of internal checks on the analysis.

As regards sensitivity, we must first divide the elements into three classes: (i) major elements, $> 1\%$ dry weight of the tissue; (ii) essential trace elements, $< 1\%$ dry weight but essential to the metabolism; and (iii) the rest, $< 1\%$ dry weight and non-essential as regards metabolic processes. Among the first group, are many light elements, with short half-lives, and activation analysis gives poor sensitivity with them. This method would therefore not be used in their estimation.

TABLE I

ACTIVATION ANALYSIS FOR SOME LIGHT ELEMENTS

Element	*Activation Process*	*Half-life*	*Sensitivity*
B	$^{11}B(p,n)^{11}C$	20 min	10^{-9} g
C	$^{12}C(d,n)^{13}N$	10 min	$2{\cdot}5 \times 10^{-7}$ g
D	$^{2}H(\gamma,n)^{1}H$	prompt	$2{\cdot}0 \times 10^{-3}$ g
N	$^{14}N(d,n)^{15}O$	2 min	10^{-6} g
O	$^{18}O(p,n)^{18}F$	1·9 h	$5{\cdot}0 \times 10^{-5}$ g

Also included in the major elements group are a number of elements which can be estimated by activation analysis using the (n,γ) reaction. Here sensitivity is either good or excellent (Table II).

Inherent in this and the following tables are the following assumptions; (i) neutron flux, 10^{12} neutrons per cm^2 per sec; (ii), irradiation for the half-life or for one month (whichever is shorter) and, (iii), chemical processing time is allowed for. Although the sensitivity is very good,

* *Present address: Chemistry Dept., The University, Reading, Berks.*

alternative methods should always be considered; for instance flame photometry is just as sensitive for the alkali metals.

TABLE II

ACTIVATION ANALYSIS FOR BIOLOGICAL MAJOR ELEMENTS

Element	*Nuclide Produced*	*Half-life*	*Sensitivity*	*Interferences*
Ca	^{45}Ca	165 d	10^{-7} g	
Ca	$^{49}Ca/^{49}Sc$	9 min/1 h	3×10^{-6} g	S
K	^{42}K	12·5 h	10^{-9} g	
Mg	^{27}Mg	9·5 min	2×10^{-7} g	Mn
Na	^{24}Na	15 h	10^{-10} g	
P	^{32}P	14 d	10^{-10} g	S, Cl
S	^{35}S	87 d	5×10^{-5} g	Cl

If we turn now to the essential trace elements, activation analysis, again using the (n,γ) reaction, really comes into its own.

TABLE III

ACTIVATION ANALYSIS FOR ESSENTIAL TRACE ELEMENTS

Element	*Nuclide Produced*	*Half-life*	*Sensitivity*	*Interferences*
Cl	^{38}Cl	37 min	4×10^{-8} g	Br
Co	^{60}Co	5·3 yr	2×10^{-9} g	
Cu	^{64}Cu	13 h	10^{-10} g	
Fe	^{59}Fe	45 d	10^{-7} g	
Mn	^{56}Mn	2·6 h	2×10^{-11} g	Fe
Mo	$^{99}Mo/^{99}Tc$	2·8 d/6 h	10^{-8} g	U
Se	^{81}Se	19 min	5×10^{-9} g	Br
Zn	^{69}Zn	14 h	2×10^{-8} g	

For all the elements in this table, with the exception of iron, activation analysis offers the most sensitive method of estimation. A large number of other elements found in biological tissue (i.e. non-essential elements) can also be estimated using the activation analysis (n,γ) reaction. Some data are given for fission product elements in Table IV. For some of these elements the sensitivity is increased a thousandfold over the previous best method.

The second advantage is freedom from contamination. This advantage only applies after the sample has been irradiated, and we have induced radioactivity present in it. Up to this point in the analysis, the danger of contamination is very real. In tissues containing bacteria, the bacteria can cause contamination; in cutting the sample with a stainless steel knife, the steel dust, containing cobalt and manganese, can give wrong results; in the analysis of blood, to which the anti-coagulant heparin has been added, the heparin can modify trace element concentrations quite markedly.

TABLE IV

ACTIVATION ANALYSIS FOR ESSENTIAL TRACE ELEMENTS

Element	*Nuclide Produced*	*Half-life*	*Sensitivity*	*Interferences*
Ba	^{139}Ba	1·4 h	2×10^{-9} g	La, U
Ce	^{141}Ce	32 d	10^{-9} g	Pr, U
Cs	^{134m}Cs	3·2 h	2×10^{-9} g	Ba, U
I	^{128}I	25 m	2×10^{-9} g	Br, Cl ,U
Ru	^{105}Ru	4·5 h	2×10^{-9} g	Os, U
Sr	^{87m}Sr	2·8 h	3×10^{-8} g	U
Y	^{90}Y	2·5 d	10^{-10} g	U, Zr
Zr	^{97}Zr	17 h	10^{-7} g	U

The third and very important advantage is that we have several parameters to work with—half-life, γ-ray energies, β-radiation energies —and unless all of these are correct for the nuclide we are attempting to estimate, we cannot have the correct nuclide. An example will make this point clear—suppose we have carried out a chemical separation in order to estimate the gallium content of a sample. If we count the β-radiation over a period of time, we can estimate the half-life of the nuclide we are counting. If this half-life is not in close agreement with the half-life of the gallium nuclide in question, then it cannot be the expected nuclide, and must be one which the chemical separation has not successfully removed.

Another possibility, which has only recently been recognized, is the use of inactive isotopes, followed by activation analysis. A good example of this technique is in the medical field—in the iron metabolism of pregnant women. Since is was not desirable to give these patients radioactive iron, what was done was to give iron-58, a rare stable isotope of iron. After a suitable period of time, a blood sample was removed, and irradiated with neutrons. The iron-58 undergoes a (n,γ) reaction to give iron-59, and can readily be detected at low enrichments.

Turning now to disadvantages, the first of these is the collection of samples to be irradiated. This is often a long and laborious job. Secondly, the method is limited to elementary analysis, and takes no account of the chemical state of the element we are estimating. Finally, and somewhat paradoxically, there is the poor sensitivity of the method. The ultimate aim of the biochemist and biologist is to know the number of tracer atoms in any particular cell, and to do this we would be required to estimate each element in the region of 10^{-12}–10^{-18} g. This cannot be done, and so activation analysis, in a certain sense, is not sensitive enough (but see Table V).

TABLE V

RARE ELEMENTS IN BLOOD CELLS

Element	*μg/ml cells*	*atoms/cell*	*Reference*
Al	0·064	110,000	Kehoe *et al.* (1940)
Au	0·00004	10	Brune *et al.* (1961)
Cr	0·020	18,000	Herring *et al.* (1960)
Pb	0·34	77,000	Willoughby *et al.* (1938); Kehoe *et al.* (1940)
Mn	0·022	19,000	Bethard *et al.* (1962)
Hg	0·0067	1,600	Comar *et al.* (1962)
Ni	0·049	40,000	Herring *et al.* (1960)
U	0·00055	110	Boirie *et al.* (1962)

After the irradiation of a tissue, there is a large amount of β- and γ-radiation emitted. After the decay of the short-lived material, the β-radiation is mainly due to phosphorus-32. Six months later, after the decay of the phosphorus-32, the main isotopes emitting β-radiation are sulphur-35 and calcium-45. The main initial γ-activity is due to sodium-24, chlorine-38, potassium-42 and manganese-56. After a week these have decayed and the long-lived γ-activities become prominent, in particular those of iron-59, zinc-65, and rubidium-86. These isotopes are common to all animal and vegetable tissues, but it is interesting to note that certain trace elements are found in higher concentrations in certain tissues. There are numerous instances of this—muscle tissue concentrates rubidium, caesium, and the halogens; the kidney concentrates cadmium and mercury, the choroid of the eye concentrates barium, and the thyroid gland concentrates iodine. There is also a possibility that the thyroid concentrates vanadium, but further work will have to be done to establish this with certainty.

Certain tissues are difficult to work with, and none more so than bone, since we cannot grind it into a powder because of the danger of contamination. Another difficulty with this tissue is that we require a chemical separation to remove phosphorus before we can look for the trace elements. Phosphorus is usually precipitated as calcium phosphate, and since this adsorbs many trace elements, it adds to our difficulties.

Another tissue of great interest is blood. An immediate problem here is the difficulty of sampling, since the needle of a hypodermic syringe is made of metal and is a potential source of contamination. This has been overcome by using nylon or polythene-coated syringes; up to now, the estimation of some sixty elements in blood has been carried out. Some recent analyses are given in Table V. (See also Butt *et al.*, 1964; Brune *et al.*, 1964; Gofmann *et al.*, 1962; Natelson *et al.*, 1962; Niedermayer *et al.*, 1962; and Parr and Taylor, 1964).

The levels vary surprisingly little; this fact is somewhat unexpected since we might expect the levels to vary with such factors as health, food intake, and place of occupation. Finally, it is interesting to compare the values obtained by activation analysis and other methods for trace elements in blood samples. The appropriate data are given for manganese in Table 6: only activation analysis gives consistent results (Bowen, 1962).

TABLE VI

μG MN/ML WHOLE BLOOD

Activation Analysis	*Colorimetry*	*Spectrometry*	*Flame Photometry*	*Microbiological Assay*
0·024 (1956)	0·020 (1912)	0·12 (1940)	1·03 (1957a)	0·065 (1947)
0·029 (1961)	0·098 (1920)	0·0016 (1948)	0·46 (1957b)	
0·014 (1961)	0·58 (1955)	0·48 (1956)		
0·012 (1961)	0·15 (1958)	0·010 (1957)		
0·011 (1962)	0·32 (1961)	0·044 (1960)		
0·011 (1963)	0·15 (1962)	0·20 (1962)		

Activation analysis is a technique which can be combined well with other, more standard, techniques—in particular, with chromatography. In most cases, the spots of a paper chromatogram can be developed with a suitable reagent or examined by ultra-violet light. If this is not the case, then an easy way to find the spots is to irradiate the entire chromatogram. The spots will then emit much more radiation than the filter-paper background. The method is especially suitable for phosphorus compounds which do not normally develop easily.

In conclusion, it is interesting to see just what work has been done in this field. As can readily be appreciated from the table, a great deal remains to be done.

TABLE VII

USE OF ACTIVATION ANALYSIS IN BIOCHEMISTRY

No. of papers	*Elements concerned*
> 20	As, Au, Cu, Mn, Na, P, Sr, Zn
10–19	Ba, Br, Cl, Co, Cs, I, K, Se
5–9	Ca, Cd, Eu, Fe, Hg, La, Mg, Mo, Rb, S, Sb, Sm, V, W
2–4	Al, B, Ce, Cr, Dy, Er, Ga, Gd, Hf, Ho, Li, Lu, Nd, Os Pr, Sc, Sn, Tb, Ti, Tm, U, Yb
1	Ag, In, Pt, Re, Si, Te, Th, Tl, Y, Zr

REFERENCES

1. Willoughby, C. E. and Wilkins, E. S. (1938). *J. Biol. Chem.* **124**, 639.
2. Kehoe, R. A., Cholak, J. and Story, R. V. (1940). *J. Nutrit.* **19**, 579.
3. Herring, W. B., Leavell, B. S., Paixao, L. M. and Yoe, J. H. (1960). *Amer. J. Clin. Nutrit.* **8**, 846.
4. Bethard, W. F., Schmitt, R. A. and Olehy, D. A. (1962). General Atomic Report, GA-2803.
5. Brune, D., Frykberg, B., Samsahl, K. and Wester, P. O. (1961). Aktiebolaget Atomenergi Report AE 60.
6. Comar, D., Le Poec, C., Joly, M. and Kellershohn, C. (1962). *Bull. Soc. Chim. France*, 56.
7. Bowen, H. J. M. (1963). Report AERE-R 4196. H.M.S.O.
8. Boirie, C., Bosc D., Hugot, G. and Platzer, R. (1962). "Determination of U in blood," *Acta Chem. Acad. Sci. Hung.* **33**, 281.
9. Gofmann, J. W. *et al.* (1962). *Advanc. Biol., Med. Phys.* **8**, 1.
10. Natelson, S., Leighton, D. R. and Calas, C. (1962). *Microchem. J.* **6**, 579.
11. Brune, D., Samsahl, K. and Wester, P. O. (1963). *Atompraxis,* **9**, 368; (1964). Aktiebolaget Atomenergi Report AE 134.
12. Niedermayer, W., Greitz, E. E. and Holley, H. L. (1962). *Arthritis Rheum.* **5**, 439.
13. Butt, E. M. *et al.* (1964). *Arch. Env. Health,* **8**, 52.
14. Parr, R. M. and Taylor, D. M. (1964). *Biochem. J.* **91**, 424.

Chapter 24

THE PRODUCTION OF A HOMOGENEOUS BIOLOGICAL MATERIAL FOR INTERCOMPARISON OF ELEMENTARY ANALYSES BY DIFFERENT LABORATORIES

H. J. M. BOWEN* AND P. A. CAWSE

Isotope Research Division, Wantage Research Laboratory, (A.E.R.E.), Wantage, Berks., England

INTRODUCTION

It has been pointed out by Bowen[1] that there are few standard biological materials available. Such materials would be useful for the intercomparison of analyses from different laboratories.

Any substance chosen as a biological standard must fulfil several conditions. First, it must be available in amounts of the order of hundreds of kilograms. Secondly, the material must be easily processed to form a homogeneous mixture. Thirdly, the material must be capable of prolonged storage without decomposition.

In this work leaf tissue from Marrow Stem Kale (*Brassica oleracea*) was chosen.

PRODUCTION OF TISSUE

About 0·2 hectares were sown with 10 kg kale seed in rows 50 cm apart. No spraying or dusting was needed. Approximately 1·6 tons of fresh leaf were collected.

DRYING

Fresh kale leaf contains about 90% of water, most of which was removed by air drying at room temperature. The rest of the drying process was carried out in ovens at 105° C. The dry tissue was stored in large polythene bags until 100 kg had been obtained.

PRELIMINARY CRUSHING AND SEPARATION OF VASCULAR TISSUE

The dried leaves were passed through a pair of 15-cm oak rollers. This left tough lengths of vascular tissue (5%) intact but crushed the lamina (95%) which was sieved out using a polythene sieve with holes of 6·35 mm diameter.

GRINDING

A grinding mill with Perspex working surfaces was used. The ground tissue was blown out of the mill by the draught of the rotor and passed

* *Present address: Chemistry Dept., The University, Reading, Berks.*

through a Perspex sieve plate with 1·6 mm diameter holes into a polythene storage bin of capacity 0·1 m^3 which could hold up to 40 kg of ground powder. The bin was air-tight apart from a small attached filter of crumpled terylene net, which allowed air (but not dust) to escape.

Sieving

The leaf powder was sieved through a nylon sieve with an aperture size of 220 μ. The coarse fraction corresponding to 37% of the dried leaf was rejected. The final recovery of fine powder, 91 kg, was 56% of the starting material, 2% of the tissue being lost by spillage.

Mixing

Mixing was carried out using a polythene shovel.

Storage

The powdered kale is stored in sealed polythene bins at 2° C. Bacteriological observation showed that it was practically sterile.

Some Properties of the Leaf Powder

The pale green powder consists of irregular discs of tissue with a wide range in sizes from 220 μ downwards and has a bulk density of about 0·56 g cm^{-2}. The distribution of particle sizes is shown in Table I.

TABLE I

Distribution of Particle Sizes in Kale Powder

Sieve fraction	*Particle size range*	*% of tissue*
> 100 mesh	150–220 μ	15·6
100–200 mesh	74–150 μ	32
200–300 mesh	53–74 μ	13·6
< 300 mesh	< 53 μ	38·4

The water content varies between 4·5 and 5·1% depending on local humidity and exposure time.

The ash content was measured and found to be 17·5%.

Testing Homogeneity

The homogeneity of the bulk powder was tested firstly by non-destructive methods. The natural beta activity, mainly due to potassium-40, was used in one series of tests. Analysis showed that the variation in count rate could be ascribed equally well to random statistical fluctuations as to inhomogeneities in the powder.

The effects of taking samples of different sizes for destructive analysis were tested in several ways. In one series of tests, the powder was extracted with boiling acetone and the weight loss recorded. The optical densities of the extracts were compared at the absorption maximum for chlorophyll-A.

Fifteen 1-g and fifteen 0·1-g samples were randomly chosen. They were exposed to boiling acetone in Soxhlet thimbles for 90 min. The residues were dried, cooled and weighed. The mean losses in weight per g of sample were $0{\cdot}1345 \pm 0{\cdot}0080$ g and $0{\cdot}1336 \pm 0{\cdot}0135$ g for the 1-g and 0·1-g samples respectively.

The extracts were made up to 250 ml and 25 ml respectively and the optical density read off at 665 mμ using 1-cm silica cells. The mean optical densities per g of powder were $0{\cdot}2666 \pm 0{\cdot}0080$ and $0{\cdot}2584 \pm 0{\cdot}0133$.

In another series of tests the powder was wet-ashed, and sodium and potassium determined by flame photometry. The results are shown in Table II.

TABLE II

SODIUM AND POTASSIUM ANALYSES ON KALE SAMPLES (15 REPLICATES)

Sample weight	*Mean p.p.m.* Na	*Mean p.p.m.* K
1·0 g	2759 ± 41	29000 ± 750
0·1 g	2639 ± 66	28162 ± 1443
0·01 g	3296 ± 477	25500 ± 1480

A more precise test of homogeneity was provided by ashing fifteen 1-g and fifteen 0·1-g samples of kale powder. The ash contents were found to be $17{\cdot}68 \pm 0{\cdot}04\%$ and $17{\cdot}44 \pm 0{\cdot}23\%$ respectively. The variance between the whole group was greater than the variance between the sub-groups.

These results can be summed up as follows:

The kale powder appears to be homogeneous. Samples as small as 0·01 g appear to be representative, but larger samples of 0·1 g–1 g are recommended to increase analytical precision.

REFERENCE

1. Bowen, H. J. M. (1964). *L'analyse par radioactivation et ses applications aux sciences biologiques*. Vendôme: Presses Universitaires de France, p. 199.

Chapter 25

TRACER TECHNIQUES IN ANALYSIS

J. E. WHITLEY

Scottish Research Reactor Centre,
East Kilbride, Near Glasgow, Scotland

Most people using activation analysis are interested in the determination of trace quantities of material. There are, however, other methods of analysis in which advantage is taken of the extremely small quantities of radioactive material which can be detected by the radiation emitted. The most useful of such techniques is Isotope Dilution Analysis. The use of radioactive tracers as indicators, and in radio-release analysis is not so widespread, but worthy of consideration, since these methods may be particularly useful in certain circumstances. These methods may offer an alternative method of analysis for a particular element where the activation method is not sufficiently sensitive.

In principle the isotopic dilution method is very simple, and may be employed with inactive, as well as radioactive, isotopes. The former requires measurements in a mass-spectrometer, whilst the latter requires relatively simple radiation detection equipment. Only the radioactive isotope method will be discussed here.

To a weighed amount of sample is added a weighed amount of the compound or element to be determined, radioactively labelled. The labelled element may be a simple radioactive isotope of the element to be determined, or, for organic analysis, the labelled substance could be an organic chemical in which some of the normal carbon-12 atoms have been replaced by radioactive atoms of carbon-14.

The sample is then treated to ensure homogeneity of the mixture of labelled and inactive compound, and the compound separated again, purified, and the specific activity of the isolated labelled compound is compared with that of the original. The labelled compound recovered will have been diluted by that in the sample, and its consequent decrease in specific activity will provide a measure of the amount of the compound in the sample. Specific activities may be expressed in units of radioactivity, or arbitrary units such as counts per minute per milligram, provided that both the original labelled compound, and that recovered, are chemically identical. If the compound is isolated from the mixture as a derivative, the change in molecular weight must be taken into account by expressing the specific activity in units of radioactivity per millimole.

The separated compound must be pure, although the separation need not be quantitative, provided that sufficient for a specific activity determination is recovered. This feature makes the method particularly useful for the determination of small quantities of organic materials which are difficult to process chemically.

Consider a sample containing W g of the compound or element to be determined. To it is added X g of the labelled material of specific activity S_1. If the specific activity of the separated and purified material is S_2, then, since the total activity is unchanged by the mixing, the specific activities can be inter-related as follows:

$$\text{Total activity} = S_1 \times X = S_2(W+X)$$

then

$$W = X\left(\frac{S_1}{S_2} - 1\right)$$

When the specific activity of the tracer is very high, so that the weight of the labelled material added is negligible compared with that of the inactive material in the sample, a situation frequently encountered in inorganic analysis, a simplification is possible. The steps in an analysis are essentially the same, except that the measurements required are of the total radioactivity added, A_1, and the total radioactivity recovered, A_2, together with the weight of recovered material, W_2. Then the weight of the material in the original sample,

$$W_1 = W_2 \times \frac{A_1}{A_2}$$

The radioactivity measurements may be expressed in counts per minute and the weight of constituent determined by any convenient method, e.g. titration or spectrophotometry. The problems involved in the utilization of the above methods are best illustrated by reference to particular examples.

In "classical" methods of isotope dilution, we make use of the equation,

$$W = X\left(\frac{S_1}{S_2} - 1\right)$$

A typical example is the determination of naphthalene in coal tar products.[1] Coal tar is such a complex mixture that separation of the naphthalene must precede its determination. The standard method for the separation of the naphthalene was by distillation, and collection of the fraction distilling in a given temperature range. The determination was completed by drying and weighing the naphthalene. This method depended on reproducible, 100%, yields in the distillation and drying procedures, which could not be guaranteed. McDonald and Turner[2] made a number of analyses by the isotopic dilution method. In each a known weight of carbon-14 labelled naphthalene was added to a known weight of tar and the naphthalene separated by the method outlined above, and carefully purified. The specific activities of the tracer, and recovered naphthalene were then compared by liquid scintillation counting. A routine determination by the standard method gave 5% pure naphthalene in the tar compared with 8·30% by dilution analysis. In these results, which are typical, the losses in the separation of the naphthalene are of no consequence in the dilution analysis, although they are responsible for the low results of the standard method.

It is possible in most inorganic applications, except for the determination of traces, to take advantage of the simplification,

$$W_1 = W_2\left(\frac{A_1}{A_2}\right)$$

This method may be used with advantage to check a separation, by providing a measure of losses of the constituent of interest due to solubility or coprecipitation, or by use of appropriate tracers to identify other elements, which cause positive errors owing to coprecipitation. It also permits the use of methods known to be non-quantitative, but having advantage over those normally employed, for example, in the purity of the precipitate. The historical method for the determination of lead in rocks, as used by Hevesey and Hobbie,[3] is a good illustration of this method. A small amount of Radium-D (a naturally occurring radioactive isotope of lead) was added to a solution of a rock sample. After a chemical separation of the lead, involving carrier techniques for the very small quantities present, the lead was finally deposited electrolytically and weighed. The radioactivity of the deposit was measured, and compared with that added, to determine the efficiency of the recovery, which was found to vary from 53–100%. Without the correction provided by the tracer, the results would have been seriously in error.

When the constituent to be determined is present only in trace quantities, i.e. when W is very small, the difference between S_1 and S_2

may be of the same order as the possible errors in their measurement, unless X, the weight of the added tracer is also made very small. This is not normally convenient, owing to the difficulties associated with handling very small quantities of material. These difficulties have been avoided in a recently developed method [4] which requires the recovery of equal but minute and unmeasured amounts of the element, both from the original tracer and from the sample with added tracer, followed by determination of the ratio of the radioactivity in the two fractions. The principle of this technique is also applicable to radiochemical separations in activation analysis. It can be made clear by illustration; (i) in precipitation reactions, *equal* amounts of precipitate can be obtained from the tracer solution and sample + tracer solution by adding *in*sufficient precipitating reagent in each case; (ii) in solvent extraction, advantage can be taken of complex formation using a *deficiency* of the complexing reagent to extract the *same weight* of compound from both solutions. The principle can obviously be extended to other separation techniques, such as ion-exchange and electrodeposition. For some analyses, this method offers an attractive alternative to radioactivation for the determination of trace amounts of material.

The advantages of the isotopic dilution method are: (i) the activity handled is negligible; (ii) it is not necessary to irradiate the sample; and (iii) in some cases (if a carrier-free isotope or one of high specific activity is available) a higher sensitivity can be reached than by activation analysis. The disadvantages include those common to other conventional methods of destructive analysis, such as the introduction of impurities from reagents. To facilitate development of this substoichiometric technique, Ruzicka and Stary have recently suggested separation procedures for the determination of 24 metals.[5]

An alternative method for the determination of trace quantities is the use of radioactive reagents. Whilst this method can also be applied to micro-scale determination, it is used to its greatest advantage for trace determinations. The outline of the method is as follows: (i) treat a weighed amount of sample with a radioactive reagent to convert the trace constituent quantitatively to a labelled derivative; (ii) remove excess of reagent, e.g. by decomposition; (iii) add a large known weight of inactive derivative (large enough for weight of active derivative to be ignored; (iv) ensure equilibration between active and inactive derivatives; (v) separate derivative and purify; (vi) determine the specific activities of reagent and derivative. Mathematically this becomes,

$$W = W^1 \frac{S_2}{S_1}$$

where W millimoles is the amount of trace constituent
W^1 millimoles is the amount of inactive derivative added
S_1 the specific activity of added reagent
S_2 the specific activity of recovered derivative.

This method is not universally applicable, since it depends on the availability of a labelled reagent which forms a derivative quantitatively with the trace constituent.

In ion-exchange separations, chromatography, and electrophoresis, it is frequently necessary to identify the separated fractions. This is straightforward enough when the fraction of interest is coloured, or readily forms a coloured derivative. Where this is not the case, the labelling of the substance with radioactivity renders it easily detectable by the radiation emitted. All that is necessary is to add the appropriate tracer to the solution and ensure that equilibrium is established before applying it to the paper or the column. In column separations, the easiest method of detecting the radioactive fraction is to allow the eluate to pass through a flow type radiation detector as it leaves the column. Location of the zones on a paper chromatogram can be carried out by scanning with a detector, or by autoradiography.

Turning now to radio-release analysis, the radioactivity of material displaced by the substance to be determined is used to measure the initial quantity of replacing substance. This method is particularly suitable for the determination of traces of ionic substances using an ion-exchange column. A suitable radioactive substance is deposited on an ion-exchange column. When the sample solution is passed slowly through the column under the proper conditions, the ion being determined displaces the radioactive species stoichiometrically, and this may be determined by counting the radioactivity in the solution from the column. This method has been used in determinations of dissolved oxygen in water. The water sample was passed through a column packed with copper turnings, on which metallic thallium-204 was deposited. The oxygen in the sample converts the metal to the thallous salt quantitatively, and this is washed off and counted. 0·2 p.p.m. of oxygen could be determined by this method. Another application is the determination of fluoride ion concentration in aqueous solutions.[6] Under highly selective conditions tantalum-182 is extracted from sulphuric acid solutions as a fluoride complex into diisobutylketone. The amount of radioactivity extracted is proportional to the fluoride ion concentration. A sensitivity of 40 μg fluoride ion is quoted for the conditions of the experiment.

In order to obtain the maximum benefit from techniques based on any of the principles outlined, one should always be aware of possible sources of error, and how they can affect the analytical results. Since most sources of error are common to all of the techniques described, they can be considered together. First, there is the isotope effect. This effect, ascribed to different isotopic masses for an element, can appear as a difference in reaction velocity for isotopes of the same element. For example, a carbon–hydrogen bond is more readily ruptured than a carbon-tritium bond. Mathematical and experimental investigations have shown that only among the lighter elements is the isotope effect significant in analysis, and having regard to the accuracy obtainable in radioactivity measurements, it may be ignored except for tritium. Even then, if it is possible to ensure that the tritium remains firmly bonded within the molecule, and so acts, in effect, as a tracer for some other grouping, the difference in reaction rate can be ignored.

It is important that the added radioactive material, and the material to which it is added, be in exactly the same chemical form, before performing any chemical manipulations, in order that they behave identically. It is useless for the radioactive isotope to be present as a complex, whilst the element to be traced is in a simple ionic form, or for the radioactive isotope to be in a different valency state from the element to be traced.

The tracer must be radiochemically pure as well as chemically pure. Organic substances labelled with radioactive isotopes can be degraded by the radiation emitted with the labelling atoms. This effect is only important for substances of high specific activity.

Finally, there is the problem of measurement of the radioactivity. If α- or β-radiation is emitted, the absorption of the radiation in the sources themselves can be significant, and so this effect should be corrected for, or the sources ensured to be of identical dimensions. Owing to the statistical nature of radioactive decay, a given measurement may differ from the true value by an amount governed by the total number of disintegrations recorded. Increased accuracy can only be achieved by increasing the number of disintegrations recorded, by increasing the duration of the measurement, or by increasing the amount of radioactivity in the sample. The accuracy of the final result can be no greater than the precision of the measurement of the radioactivity, and this has been shown to be from ± 2 to $\pm 5\%$.

References

1. McDonald, W. S. and Turner, H. S. (1952). *Chem. and Ind.*, 1001.
2. McDonald, W. S. and Turner, H. S. (1964). *Chem. Trade J. & Chem. Eng.* March 6, p. 323.
3. Hevesey, G. and Hobbie, R. (1932). *Z. Anal. Chem.* **88**, 1.

4. Ruzicka, J. and Stary, J. (1963). *Talanta*, **10**, 287.
5. Ruzicka, J. and Stary, J. (1964). *Talanta*, **11**, 697.
6. Moore, F. L. (1963). *Anal. Chem.* **35**, 1032.

Further Reading

1. Lambie, D. A. (1964). *Techniques for the Uses of Radioisotopes in Analysis.* E. and M. N. Spon Ltd.
2. "Radioactive Isotope Dilution Analysis" (1964). *R.C.C. Review* 2, Available from the Radiochemical Centre, Amersham, Bucks.

EXPERIMENTAL SECTION

I

SETTING UP A GEIGER COUNTER

A Geiger counter is used in several of the scheduled experiments. The following notes are provided to safeguard the Geiger and ensure its correct use. (See also Chapter 4).

The counter consists of a thin central wire, the anode, surrounded by a metal cylinder, the cathode. These two electrodes are enclosed in a glass case, one end of which consists of a thin window of aluminium or mica. The counter is filled with a gas, such as argon, which is normally non-conducting.

When an ionizing particle enters the counter, it causes the gas to conduct electricity. If a voltage is applied between anode and cathode, a charge will be collected by the anode which can be amplified and counted. The voltage required between anode and cathode to ensure collection of the charge varies between 400 and 1600 V, depending on the constitution of the gas in the counter. The operating voltage is written on the body of the counter, and it is unwise to exceed this voltage by more than about 50 V.

All the counters used are designed for the detection of β-particles; however, if γ-rays are also present they will be detected but with less efficiency.

As the voltage between anode and cathode is increased, the count rate from the source increases steadily until it reaches a value which is approximately constant over a range of 50–100 V. The variation in the counting rate in this region should not be more than 1% per 100 V for a good counter. If the voltage is increased still further the counting rate increases rapidly and eventually the counter will be destroyed.

Before switching on, make sure that both high voltage knobs are fully anti-clockwise. Place a source under the counter and increase the voltage slowly until counts are recorded. Plot a graph of count rate vs. V for intervals of 10 V until the end of the "plateau" is reached. Determine the percentage increase in count rate per volt in the plateau region. Choose the operating condition at the midpoint of the plateau. This position has the advantage that small fluctuations in voltage will not affect the count rate.

2

ABSORPTION OF β-RAYS

Introduction

The energies of β-rays emitted by a radioactive decay are distributed from zero to a maximum which is characteristic of the decay. For the identification of an unknown β-emitting isotope it is often necessary to determine by experiment the maximum energy of the β-ray spectrum. The absorption of β-rays is a complicated phenomenon and there is no exact mathematical expression analogous to the exponential law describing the absorption of mono-energetic γ-rays. It is found experimentally, however, that the proportion of β-particles (produced by radioactive decay) which emerges through a given thickness of absorber decreases nearly exponentially with increasing thickness. The absorber thickness is conveniently expressed in mg cm^{-2} and the results are then almost independent of the nature of the absorber.

Apparatus

Planchette containing a source of ^{204}Tl, set of aluminium absorbers, Geiger counter, stopwatch, E.R.D. lead castle.

Procedure

The planchette should be put on the appropriate Perspex tray and placed on shelf 4 in the lead castle. The tray for the aluminium absorbers goes in on shelf 1.

The variation in counting rate with absorber thickness should be determined for all the absorbers provided and one measurement made without any absorber in. The results should be plotted on semi-log graph paper, and the following observations made from the graph:

(i) The roughly exponential nature of the absorption should be verified.

(ii) The half-thickness, which is defined as the thickness of absorber necessary to reduce the count-rate to half its original value, should be determined.

(iii) The maximum energy of the β-rays in mg cm^{-2} should be determined.

(iv) The "tail" due to secondary bremsstrahlung radiation should be identified.

In addition, the following questions should be answered:

1. It will be observed that the graph falls below the expected value for small absorber thicknesses; what is the explanation for this effect?
2. The energy in MeV of the β-rays can be calculated using Feather's equation:

$$E_{max} = 1 \cdot 84R + 0 \cdot 294$$

where E_{max} = maximum energy of β-rays
R = range in g cm^{-2} previously calculated.

Check this value against the tabulated value, e.g. in *The Radiochemical Manual*, Part 1, Physical Data; published by The Radiochemical Centre, Amersham, Buckinghamshire, 1962.

3

ANALYSIS OF A RADIOACTIVE MIXTURE USING β-RAY ABSORPTION

Introduction

The absorption of β-rays in aluminium is approximately exponential, the power of the exponential depending on the energy of the β-ray. Thus β-rays of different energies will give different straight-line gradients when plotted on semi-log graph paper. The sample given is a mixture of two isotopes which have been produced in the reactor and the object of the experiment is to determine the β-ray energies of the two isotopes.

Apparatus

Lead castle, Geiger counter, aluminium absorbers, stop-watch, scaler.

Procedure

1. Measure the count rate with 0, 150 and 2000 mg cm^{-2} absorbers.

2. Starting from zero measure the count rate for increasing absorber thicknesses, plotting the points as you go. The graph will be found to flatten off, the flat portion being due to γ-emission. Estimate the range of the higher-energy β-ray using Feather's equation:

$$E_{\text{max}} = 1{\cdot}84R + 0{\cdot}294$$

where

$$E_{\text{max}} = \text{maximum energy of } \beta\text{-rays in MeV}$$

$$R = \text{range in g cm}^{-2}$$

Extrapolate the γ-ray component to zero, subtract and replot. The resulting graph should be two straight-line segments. Estimate the energy of the lower-energy β-ray using Feather's equation. Extrapolate the higher-energy β-ray to zero, subtract and replot the points between 0 and 100 mg cm^{-2}.

3. At the end of the experiment, repeat 1.

Answer the following questions:

1. By which β-emitter is the γ-ray emitted?
2. Say what you can deduce about the half-lives of the two isotopes.
3. Using all the available information and consulting reference book of β- and γ-energies, try to decide which isotopes are present.
4. Find the half-value layer for each β-ray.

4

γ-RAY SPECTROMETRY—SINGLE-CHANNEL ANALYSER

Introduction

Just as an optical spectrometer can identify excited atoms by measuring the wavelengths of the optical radiation they emit, the γ-ray spectrometer can identify radioactive nuclei by measuring the energies of the γ-rays they emit. However, whereas the energy of typical optical radiations is in the electron-volt range, the energies of most γ-rays are in the MeV range, i.e. higher by a factor of a million or so.

The most important type of γ-ray spectrometer consists of a scintillation counter coupled through a pulse amplifier to some form of pulse height analysis equipment: in the apparatus in the laboratory this is a single channel device.

Apparatus

1. scintillation counter

This consists of a sodium iodide crystal coupled to a photomultiplier tube. γ-rays interacting with the sodium iodide crystal cause the ions in the crystal to be excited to higher energy states; on de-excitation they emit radiation, part of which is in the visible band, and the quantity of such radiation is proportional to the energy of the γ-rays. The light flashes from the crystal, as individual γ-rays interact with it, are detected by the photomultiplier tube, which delivers voltage pulses proportional in height to the γ-ray energy.

2. pulse amplifier

The pulses from the photomultiplier are a few tenths of a volt in height and are amplified in pulse amplifiers until they are in the range 5–50 V where they can be conveniently analysed by the single-channel analyser.

3. single channel analyser

This accepts pulses in the range 5–50 V and passes on to the scaler only those whose heights lie within a narrow band, which can be varied both in position and in width. By moving this band in steps from top to bottom of the voltage range and measuring the counting rate at each step a plot can be made of counting rate against pulse height, and this is known as a γ-ray spectrum. A typical spectrum for a radioactive isotope such as caesium-137, which emits γ-rays of a single energy, is shown in Fig. 44.

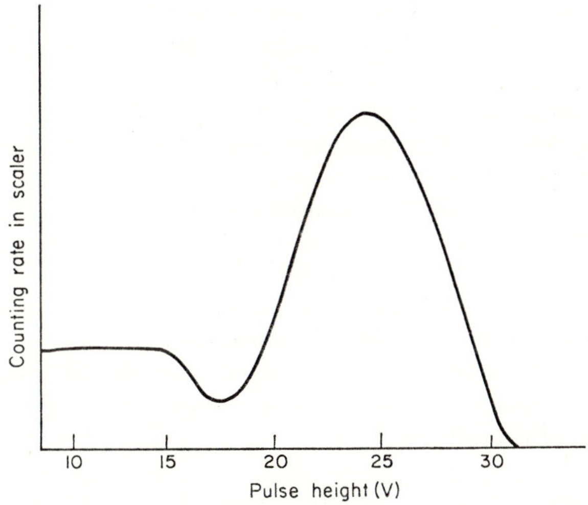

FIG. 44. Peak from full energy (0·66 MeV) of caesium-137 γ-ray.

A γ-ray spectrum from a source which emits γ-rays of several energies will exhibit a series of peaks, each of which will correspond to a γ-ray of a particular energy. The energy scale of the apparatus is approximately linear and so is easily calibrated using two or three sources of γ-rays of known energies.

4. SCALER

The scaler counts the pulses coming from the single-channel analyser; if a count is made over a known time interval the counting rate can be determined, and this is proportional to the intensity or strength of the γ-ray being analysed.

PROCEDURE

The controls on the amplifier, E.H.T. unit and scaler will normally be preset at the required position and should not require adjustment. Proceed as follows:

1. Set the single-channel analyser controls to the following positions:

Discriminator–analyser switch	—analyser
Channel-width	—1 V
Bias	—5 V on meter

2. Place one of the standard sources (^{198}Au, ^{137}Cs or ^{60}Co) on top of the scintillator counter, utilizing the spaces provided, and take a $\frac{1}{4}$-min count. Change the bias setting to 6 V and repeat the count. Repeat at 1-V intervals of bias up to 50 V or until the count has dropped to near background level (i.e. the count obtained without any source being present).
3. Subtract the background from each count and plot the γ-ray spectrum for the standard source.
4. Repeat for the other two standard sources.
5. Using the known energies of the standard sources, draw up a calibration curve relating pulse height to energy. A plot of the counter efficiency against energy can also be constructed using the known strengths of the standard sources.
6. Repeat 1–3 for the unknown source, and use the calibration and efficiency curves obtained in 5 to estimate the γ-ray energies and strength of the source.

Standard source	*γ-Ray energies MeV*	*Strength in μC*
Gold-198	0·411	0·46
Caesium-137	0·661	3·702
Cobalt-60	1·17 and 1·33	2·04

5

CRYSTAL RESOLUTION

INTRODUCTION

The resolution, R, of a scintillation counting system is defined by $R = \Delta E_n/E_n$ where ΔE_n is the full width of the photopeak at one half of its maximum height, and E_n is the mean photopeak energy. By determining the resolution of a sodium iodide crystal for different incident γ-ray energies, the variation of resolution with γ-ray energy can be found and interpreted in terms of the statistical nature of the interaction of γ-rays with matter. The two experiments A and B, performed with the available counting equipment are given below.

EXPERIMENT A

OBJECT

To determine the resolution of a scintillation counter using a single channel pulse height analyser.

APPARATUS

2 in. × 2 in. (diameter × length) sodium iodide crystal coupled to a photo-multiplier tube (the assembly being housed in a lead castle) and a single channel pulse height analyser. The last of these is an instrument which accepts pulses in the range 5–50 V and passes on to the scaler only those pulses whose heights lie within a narrow band which can be varied both in position and in width. By moving this band in steps through the voltage range and measuring the count rate at each step a plot may be made of counting rate versus pulse height. This plot is known as a γ-ray spectrum. The energy scale of the apparatus is approximately linear and this may be verified using two or three sources of γ-rays of known energy.

PROCEDURE

1. Set the single channel analyser controls to the following positions:

Discriminator–analyser switch	—analyser
Channel-width	—1 V
Bias	—5 V on meter

2. Place one of the standard sources (^{198}Au, ^{137}Cs or ^{60}Co) on top of the scintillation counter and take a $\frac{1}{4}$-min count. Change bias setting to 5 V and repeat the count. Repeat at 1-V intervals of bias up to 50 V or until the count has dropped to near background level (i.e. the count obtained without any source being present).

3. Subtract background from each count and plot the γ-ray spectrum for each source.

4. Using the known energies of the standard sources (gold-198 $-0{\cdot}411$ MeV, caesium-137 $-0{\cdot}661$ MeV and cobalt 60 $-1{\cdot}17$ MeV and 1·33 MeV) draw a calibration curve relating pulse height to energy.

5. Determine for each photopeak the full width ΔE_n at one half of the maximum height. The ratio $R = \Delta E_n/E_n$ where E_n is the mean photopeak energy, is generally called the "resolution" and sometimes the "instrumental line width". Plot a graph of R^2 against $1/E$. Interpret the result in terms of statistics.

EXPERIMENT B

Object

To determine the resolution of a counting system comprising a 3 in. × 3 in. sodium iodide crystal (thallium activated), photomultiplier tube, preamplifier and Laben kicksorter (i.e. multi-channel pulse-height analyser).

Apparatus

Lead castle, crystal and photomultiplier assembly, Laben kicksorter and printer, ^{60}Co, ^{137}Cs and ^{198}Au sources.

Procedure

Place the sources in the lead castle and count for approximately 5 min. At the end of the counting period, print out the results and plot a graph of counts versus channel number. Plot also the mean photopeak energy, E_n, against the channel number using the following data:

^{198}Au 0·411 MeV γ-rays (96%)
^{137}Cs 0·661 MeV γ-rays
^{60}Co 1·33 and 1·17 MeV γ-rays

Verify the linearity of the plot and determine the bias setting of the pulse height analyser. Interpolate non-photopeak regions through photopeaks and determine the full width ΔE of each photopeak at one half of the maximum height. The ratio $\Delta E_n/E_n = R$ is the "resolution" or the "instrumental line width". Estimate the errors due to uncertainties on the subtraction of non-photopeak regions and if time allows perform an analysis of the sources individually and determine the resolutions. Plot a graph R^2 versus reciprocal incident γ-ray energy and interpret the result in terms of statistics. Estimate, for example, a minimum value of the mean number of photo-electrons striking the first dynode for 500 keV absorbed in the crystal. Verify by calculation that in the case of a photomultiplier with a uniform stage gain of 5, the major statistical variation is in the number of electrons striking the first dynode.

References

1. Crouthamel, C. E. *Applied Gamma-Ray Spectrometry*. Pergamon Press.
2. Sharpe, J. *Nuclear Radiation Detectors*. Methuen Monograph.

6

IDENTIFICATION OF γ-ACTIVITIES BY PULSE-HEIGHT ANALYSIS

Introduction

The energies of γ-rays emitted from material which has been made radioactive by irradiation in a nuclear reactor can serve as a means of identifying the constituents of the material. A convenient method of recording the γ-ray spectrum of such material is to use a multi-channel pulse-height analyser. As an example of this method of analysis a sample of material for analysis (in this case mouse muscle) will be irradiated in the rabbit system of the reactor. The irradiated sample will be placed on a γ-scintillation spectrometer. After a preset counting time the pulse-height analysis of the γ-spectrum will be automatically printed out. Copies of the analysis will be available for subsequent identification of the radioactivities induced in the irradiated material.

Apparatus

Sodium iodide scintillation spectrometer, Laben multi-channel pulse-height analyser with typewriter print-out.

Procedure

Plot a graph of the number of counts recorded in each channel against the channel number. (The spectrum is usually divided into 128 channels. The first channel is not a point on the γ-spectrum, but indicates the live time of the counting period.) The energies of the unknown γ-peaks are identified by comparison with the location of peaks of known γ-energies as follows: a number of known γ-sources will be counted on the same scintillation spectrometer, and the location of their γ-peaks indicated. From these details, a calibration line can be drawn of the original graph, and the energies of the unknown γ-emitters located.

A typical result, with calibration curve is shown in Fig. 45. The energy of a γ-peak in the irradiated material is identified by drawing a vertical line from the peak to the calibration line, and reading off the energy from the calibration at the right-hand side of the graph. The energies of the unknown peaks should be tabulated, and these peaks attributed to specific isotopes by consulting published tables of γ-spectra, e.g. *Gamma-ray Spectrum Catalogue*, R. L. Heath and *Gamma-ray Spectrometry*, C. E. Crouthamel. For convenience, the γ-energies of some commonly used calibration sources are listed below:

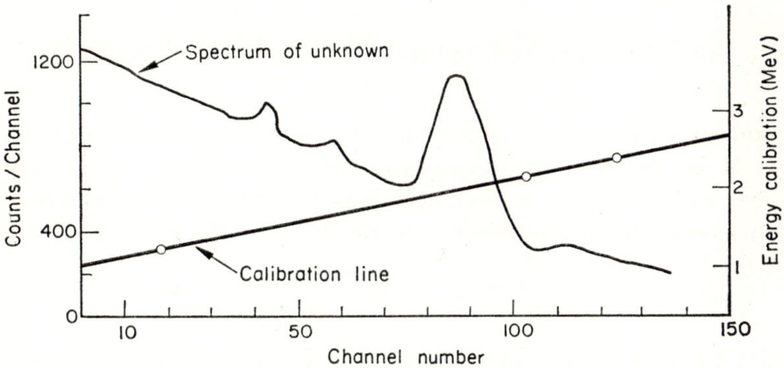

FIG. 45. Identification of γ-activities by pulse height analysis.

	Energy, MeV
^{60}Co	1·17, 1·33
^{137}Cs	0·663
^{144}Ce	0·133

7

ANTIMONY–BERYLLIUM NEUTRON SOURCE

This source is in the form of a cylinder, 3·6 cm long and 2·3 cm in diameter. It is prepared by hot compacting of antimony and beryllium powders, which are then sealed into an aluminium can. The whole source is then irradiated in the reactor to produce ^{124}Sb, a β- and γ-emitter ($t_{1/2} = 60$ d) by the reaction:

$$^{123}\text{Sb}(n,\gamma)\,^{124}\text{Sb}$$

The 1·71 MeV γ-radiation from ^{124}Sb produces a (γ,n) reaction on beryllium.

$$^{9}\text{Be}(\gamma,n)\,^{8}\text{Be} \longrightarrow 2\,^{4}\text{He}$$

The threshold energy for this reaction is close to 1·71 MeV; thus the neutrons are generated with low energies (about 25 to 30 keV) and are reduced to thermal energies by passage through 3 or 4 cm of water.

The neutron output is about $1{\cdot}5 \times 10^6$ n sec^{-1} curie^{-1} of ^{124}Sb, falling naturally with decay of the ^{124}Sb.

The γ-ray background necessitates efficient shielding. 1 mc of ^{124}Sb produces a γ-ray dose rate of 1·0 mr h^{-1} at a distance of 1 m, and the average source has a content of 1 curie. The half-value layer of the γ-radiation is approximately 1·2 cm.

Spend the minimum amount of time in the neutron house and do *not* look directly into the tank.

8

COUNTING-LOSSES: DETERMINATION BY EXTRAPOLATION OF A KNOWN DECAY CURVE

Apparatus

EtI, *n*-source, KI solution (40 mg/100 ml of I^-), 0·1N $AgNO_3$, separating funnel, centrifuge, pipette, etc., end window Geiger counter, scaler.

Object

In this experiment a source of high specific activity is prepared by a neutron-irradiation of ethyl iodide, followed by extraction of the iodine liberated in the Szilard–Chalmers process. AgI is prepared from the source, mounted, and counted. The count rate at the start of the observations will be high and counting losses considerable. As the source decays, counting losses become smaller, and the decay curve can be extrapolated back to give the counting losses at the higher rates. The half life of ^{128}I is accurately known.

Method

(*a*) Irradiate 500 g clean dry EtI with neutrons for 2 h. During this time, practise the following procedure on an inactive sample, and then apply the method to the irradiated EtI.

(*b*) Transfer the EtI to a separating funnel, add about 25 ml aqueous KI solution (0·5 g KI/litre) and about 2 ml SO_2 solution. Shake vigorously for about 15 sec, then separate the ethyl iodide and aqueous layers.

From the aqueous layer, precipitate AgI by adding $AgNO_3$ solution and a few drops of dil. HNO_3. Centrifuge off the precipitate, transfer to a large aluminium counting tray as an acetone slurry, and dry under the infra-red lamp.

(*c*) The source should be placed as near to the counter window as possible and counted as follows:

> Count for 1 min 50 sec, then read and reset the scaler, and begin counting again after 10 sec, i.e. take a series of counts at 2-min intervals. Plot the count-rate against time from the start of the counting procedure, using semi-log paper.

(*d*) Draw a straight line, of slope equivalent to a half life of 24·99 min, through the lower points of your decay graph.

(*e*) Plot the discrepancies, i.e. the extrapolated values minus the observed values at the higher rates (expressed as a fraction of the true rate) vs. observed counting rate and deduce the dead time of the counter.

(*f*) Leave the ethyl iodide over calcium chloride for the next experiment.

Notes

(1) A quantitative yield is not required, but (in order to obtain a highly active source) a yield as high as possible consistent with rapid working should be sought.

(2) The reaction in the neutron tank is: $^{127}I(n,\gamma)^{128}I$. Both cross-section and half-life are convenient for a good yield of ^{128}I. Other activities do not interfere.

(3) If R_{true} is true counting rate
R_{obs} is the observed rate
τ the dead time

$$R_{true} = \frac{R_{obs}}{1 - R_{obs}\,\tau}$$

Thus

$$\frac{R_{obs}}{R_{true}} = 1 - R_{obs}\,\tau$$

$$\frac{R_{true} - R_{obs}}{R_{true}} = R_{obs}\,\tau$$

Thus the slope of the graph of discrepancy vs. R_{obs} is τ.

9

FLUX PLOT IN THE THERMAL COLUMN OF THE UTR-100 REACTOR

Introduction

The UTR-100 thermal column consists of a 4-ft square × 5-ft long block of graphite, which at one end abuts on to the graphite reflector region surrounding the reactor core and, at the other, is closed off by a massive concrete door which may be moved back to expose the outer face of the column. Graphite is a good moderator, with a relatively small absorption cross-section and correspondingly large diffusion length and therefore fast neutrons entering the core end of the column will be efficiently slowed down with small absorption losses. Because of the large diffusion length (54 cm) the thermal neutron flux falls off relatively slowly along the length of the column, and hence the outer half of the column serves as a fairly high flux source of extremely well thermalized neutrons, which are useful in some types of experiment.

In this experiment we will measure the thermal-neutron flux at various positions in the thermal column by a foil-activation technique, and from the measured flux profile along the length of the column estimate the thermal diffusion length in graphite. If time permits, cadmium ratios will also be measured at several points (see below).

The thermal-neutron flux at any point in a nuclear reactor can be computed from the activation of a thin foil of an element such as gold, which has a conveniently large absorption cross-section for thermal neutrons. On irradiation the radioisotope ^{198}Au is formed, which is a β–γ-emitter of half-life 2·7 days. The measured disintegration rate in the foil is related to the thermal-neutron flux through the equation:

$$\Phi = \frac{A}{\sigma_n(1-e^{-\lambda t_1})\,e^{-\lambda t_2}}$$

where A = disintegration rate of ^{198}Au in the foil in d.p.s.
σ = thermal-neutron absorption cross section of Au, can be taken as 98 barns (1 barn = 10^{-24} cm^2)
n = number of Au atoms in the foil
λ = disintegration constant of ^{198}Au, sec^{-1} = $0{\cdot}693/T_{1/2}$, where $T_{1/2}$ = ^{198}Au half-life = 2·7 days

t_1 = duration of irradiation, seconds
t_2 = interval between end of irradiation and time foil is counted, seconds
Φ = thermal-neutron flux, neutron cm^{-2} sec^{-1}.

A fraction of the activation of a gold foil is due to neutrons in the resonance or $1/E$ slowing-down region of the spectrum, and a measurement of this fraction yields an estimate of the number of such neutrons relative to those in the thermal region. This measurement is carried out by irradiating gold foils covered with cadmium sheet (0·020 in. thick); cadmium has a very large absorption cross-section for thermal neutrons, which cuts off quite sharply above 0·4 eV, hence the cadmium-covered activation is due almost entirely to neutrons above this energy. The ratio of bare activation to cadmium-covered activation is known as the cadmium ratio, and it is easily seen that

$$\frac{\text{Thermal flux}}{\text{Resonance flux}} = \text{Cadmium ratio} - 1$$

Due to the relatively high absorption cross-section of gold, the flux at the foil is slightly lower than it would be in the absence of the foil; also, due to absorption of neutrons in the outer layers of the foil the flux at the centre of the foil is lower than at the surface. To correct for these effects the fluxes calculated according to the above formula should be multiplied by a flux depression factor of 1·05.

Apparatus

UTR-100 reactor, operated at a power of 3 W for about 30 min. Gold foils approximately $\frac{3}{8}$ in. in diameter and 0·002 in. thick. The foils are kept in envelopes bearing a reference number and the weight of the foil. Care should be taken that foils are not replaced in the wrong envelope. Foils should be handled with tweezers or forceps as far as possible. Slips of paper slightly larger than the foils should be wrapped round them to prevent their sticking to the Cellotape used to attach them to the graphite. Cadmium foils approximately $\frac{1}{2}$ in. in diameter and 0·02 in. thick. Scintillation and Geiger counters.

Procedure

1. The central stringer and two or three others in the thermal column will be removed by the demonstrator. Bare gold foils will be attached to the central stringer at 6-in. intervals along its length on its top surface. One foil will also be attached to each of the other stringers at a position along its length corresponding to one of the foils on the central stringer, so that a comparison can be made between the flux at the centre-line of

the column and that at various positions off-centre. The stringers will be replaced.

2. The reactor will be operated at a power of 3 W for about 30 min.

3. At the end of the irradiation, the foils will be recovered from the thermal column, replaced in their envelopes, and carried through to the laboratory for counting.

4. If time is available, cadmium-covered foils will be attached to three or four of the positions along the central stringer at which bare foils were irradiated, and the reactor will be run again at 3 W for 30 min.

5. Each group of students will take two or three foils initially and, after they have analysed them, will interchange with another group so that eventually everybody will have analysed all the foils.

6. The foils can be counted relative to one another on the Geiger counters, noting that since there will be a large variation in counting rates between foils, it will be necessary to use different foil-counter spacings. However, the relative counting rates of the foils can be maintained in this event by comparing the counting rates of a given foil at both spacings when a change is about to be made. Corrections will have to be made to allow for counter background and dead-time. The time at which each count is made should be noted.

7. Using the scintillation counters, the ^{198}Au γ-ray spectrum from one of the foils will be plotted, and the counting rate in the 0·41 MeV γ-ray photo-peak determined. The foil-crystal distance will be measured and, from tables, the photo-peak efficiency will be computed. The counting rate will be divided by the efficiency, multiplied by the factor 0·95 to allow for the ratio of 0·41 MeV γ-rays to total disintegrations, and the total disintegration rate in the foil will be computed. The neutron flux corresponding to this foil will be computed from the expression presented in Section 1.

8. The fluxes at other positions can now be derived from the results obtained in 6 and 7.

9. If the cadmium-covered foils have been irradiated, they will be counted as in 6 above, the cadmium ratios will be computed, and the flux value obtained in 8 corrected for resonance activation. The correction will be small (i.e. the cadmium ratio will be large) everywhere except in the inner foot or so of the column.

Results

1. Plot the flux as a function of distance from the inner face of the column, using log-linear paper. Make a similar plot of the flux along a line at right angles to the centre-line, if enough data are available.

2. Measure the slope of the linear portion of the plot along the centre-line and, from it, obtain a value of k.

3. From the expression

$$k^2 = \left(\frac{\pi}{2\bar{a}}\right)^2 + \left(\frac{\pi}{2\bar{b}}\right)^2 + \frac{1}{L^2}$$

we will obtain a value of L, the thermal-neutron diffusion length in graphite, where

$2\bar{a}$ = extrapolated width of the thermal column

$2\bar{b}$ = extrapolated height of the thermal column.

10

SATURATION AND DECAY OF A RADIOACTIVE SUBSTANCE

Introduction

The experiment is designed to demonstrate the effect of irradiation time and decay time on the activity of a radioactive isotope. From theory the activity of an isotope after irradiation time t can be expressed by the equation:

$$P_t = P_0(1-e^{-\lambda t})$$

where P_t = number of disintegrations per second after irradiation time t
P_0 = number of atoms activated per second
λ = disintegration constant

If the substance is then removed from the activating neutron flux the activity is expressed by the equation:

$$\frac{dN}{dt} = -\lambda N \quad \text{or} \quad N = N_0 e^{-\lambda t}$$

where N = number of activated atoms present after time t
N_0 = number of activated atoms present when substance is removed from neutron flux.

Apparatus

Scaler, Geiger tube, lead castle, magnesium discs, stopwatch.

Procedure

Six magnesium discs are placed in the reactor via the rabbit system and irradiated at 10 W for 5, 10, 15, 20, 30 and 45 min respectively. As each disc is removed from the reactor start the stop-watch and count the sample between the second and third minute after removal. In addition count the last sample every minute for periods of 30 sec until the counting rate has reached a quarter of its original value. Plot the results for the different discs on linear graph paper and the results for the final disc on semi-log graph paper.

Results

Determine the half-life of the radioactive magnesium from the decay of the final disc counted, and determine the degree of saturation $(P_t)/(P_o)$ after an irradiation time t equivalent to the half-life, $t_{1/2}$.

11

FOIL CORRECTION FACTORS

Introduction

Accurate determinations of neutron flux are commonly made by irradiating thin foils of suitable materials in the flux to be measured, carrying out an absolute measurement of the induced activity in the foil, and calculating the flux from the known values of activity, irradiation time, cross-section, etc.

For the most common activation detectors, such as gold and indium, which have relatively high cross-sections, the above calculation would only yield accurate results if very thin (i.e. less than a few mg per cm^2) foils were used. However, in many practical applications foils of much greater thickness have to be used, and in such cases a correction has to be made for the effect of the outer region of the foil absorbing an appreciable proportion of the neutrons impinging on it and hence causing the interior of the foil to "see" a somewhat smaller flux than would otherwise be the case. Because of this self-shielding effect a thick foil will exhibit a somewhat smaller induced activity per unit mass than a thin foil irradiated in the same flux. It is the purpose of this experiment to demonstrate this effect and to obtain a self-shielding correction factor for a thick gold foil.

The measurement of the induced activity in the foil is usually the most difficult step in the foil activation technique of flux determination, and in the best work is done by 4π counting of thin foils for β- and/or γ-rays. However, in many cases sufficient accuracy can be attained with relatively thick foils counted on simple β-detectors, provided that the effects of self-absorption and reflection of β-particles in such foils are taken into account. An investigation of these effects in a thick gold foil forms part of this experiment.

Briefly the experiment involves the irradiation of a stack of thin gold foils pressed together so that they simulate a thick foil, counting the activity of this thick foil, then separating the foils and counting them individually, first bare, secondly with various thicknesses of inactive gold absorber on top, and finally with various thicknesses of gold reflector beneath. β-absorption and reflection factors are obtained, and used together with the thin foil measured activities to synthesize a value for the thick foil activity which can be compared with the actual measured value.

Apparatus

Nine gold foils, each $\frac{3}{8}$ in. diameter × 0·001 in. thick; each foil has an identification number engraved on it, and is kept in a paper envelope bearing this number and the weight of the foil in mg. Aluminium foil holder for irradiation of the foils. End-window G-M tube and scaler. Perspex foil holder for counting the foils.

Procedure

1. Five of the gold foils are packed into the aluminium foil holder and irradiated in the fast transfer facility of the reactor for 5 min at a power of 100 W. A careful note should be kept of the order in which the foils are loaded into the foil holder. If time is short the irradiation will be carried out by the staff before the laboratory period starts.
2. Stack all five irradiated foils into the counter foil holder in the same order as that in which they were irradiated, and take a count. A one minute count should be sufficient, but in any case at least 10,000 counts should be accumulated.
3. Separate the foils and count each individually. Plot the count rate from each foil as a function of position in the stack, correcting for the differing weights of foils.
4. To estimate beta absorption in the thick foil, count one of the active foils covered with one of the inactive foils, then with two, three and four of the inactive foils. Plot the count-rate as a function of the number of cover foils.
5. To estimate beta reflection in the thick foil, count one of the active foils on top of one of the inactive foils, then on top of two, three and four. Plot the count-rate as a function of the number of backing foils.
6. From the data obtained in 4 and 5, calculate beta absorption and reflection factors for each foil in the original stack.
7. For each foil in the stack, correct the observed count-rate for beta absorption and reflection. Add the corrected count-rates, and compare the result with that obtained on counting the stack of foils together, as in 2.
8. Comment on the results.

12

METHOD FOR MERCURY ANALYSIS

1. Place the samples (dried human liver) in the digestion flasks.

2. Add 10 mg mercury carrier (mercuric nitrate) as 1 ml of soln.

3. Add 2 ml mixed acid (conc. nitric: conc. sulphuric, 1:1). Heat till the nitric acid is removed, eliminating any charring by the addition of more nitric acid.

4. Transfer the digested sample with washings to a (50 ml) centrifuge tube.

5. Neutralize (almost) with 20% sodium hydroxide with congo red as the indicator.

6. Heat to boiling on the water bath, and add 2 ml 1% ascorbic acid solution.

7. Spin down the Hg precipitate, wash with water and with acetone.

8. Dry the precipitate on the water bath, and dissolve it in 0·5 ml conc. nitric acid.

9. Dilute and add 0·5 ml of 10% silver nitrate solution.

10. Add 2 ml 10% sodium iodide. Heat the solution.

11. Spin down and filter off the silver precipitate and reject it. The supernatant is collected in a clean centrifuge tube.

12. Neutralize the solution with 10% ammonium hydroxide with congo red as the indicator.

13. Add 3 ml of the copper complex (Cu ethylenediamine)*.

14. Spin the precipitate down, wash it twice with cold water, and once with iso-propanol, and transfer with iso-propanol on to a weighed steel planchette.

15. Weigh the trays to determine the recovery, and count the activity, using a scintillation detector.

STANDARD

A sample of mercury has been irradiated along with the samples and has been dissolved in nitric acid and made up to 10,000 ml. Take 1 ml of this solution (containing about 0·1 μg Hg) and process it as for a sample

* The copper complex (Cu ethylenediamine) is made as follows

1. Make 10% v/v ethylenediamine in water

2. Mix 1 part 10% (w/v) copper sulphate in 10 parts ethylenediamine solution, prepared as above.

from step (4) above, i.e. omitting digestion, but with the addition of 1 ml Hg carrier.

Calculation

The counts of the standard and samples are corrected for the dead-time of the counter and the variation in chemical recoveries. The decay correction is negligible for this isotope.

The count rate (C.R.) for the determination is found as follows:

$$\text{C.R.} = \frac{\text{Counts standard (c/m)} \times 10^4}{\text{Weight of Hg irradiated (mg)}}$$

The factor of 10^4 arises from the dilution of the standard sample.

Then the mercury content of the samples in p.p.m. is given by the following:

$$\text{Hg content (p.p.m.)} = \frac{\text{Sample count (c/m)} \times 10^6}{\text{Weight of sample (mg)} \times \text{C.R.}}$$

13

METHOD FOR ANTIMONY ANALYSIS

1. Place the sample (dried human liver) in the conical beaker.

2. Add 1 ml Sb carrier (equivalent to 20 mg Sb),* and 3 ml conc. sulphuric acid.

3. Heat till fumes of sulphur trioxide are observed, and for a further three minutes.

4. Add small portions of solid sodium nitrate till the solution turns pale yellow. Continue heating until all the nitric acid is driven off.

5. Add 0·25 g solid sodium sulphite. Remove from heat and allow to cool.

6. Add 4 ml water, and 1 ml conc. HCl.

7. Transfer the solutions to 50 ml centrifuge tubes, and dilute to 30 ml.

8. Heat on the water bath, and add 3 ml of chromous chloride (about N).

9. Spin down the precipitate and wash it twice with cold water.

10. Add 1 ml As carrier (10 mg As), seven drops hydrogen peroxide (30% w/v), and 6 ml conc. HCl. Agitate the tube to assist the solution of the precipitate and as soon as solution is complete, add 2 ml saturated ammonium hypophosphite.

11. Boil the solution for 20 min.

12. Dilute the solution, and add (in this order)—5 drops of 10% copper sulphate, 10 drops potassium ferricyanide, and 10 drops cobaltous chloride, shaking the contents of the tubes thoroughly after each addition.

13. Spin down and filter off the precipitate, collecting the supernatant in a clean centrifuge tube.

14. Heat the solution, and precipitate Sb with chromous chloride as in (8).

15. Spin down the precipitate, wash it twice with water and once with acetone and transfer it to weighed planchettes with acetone.

16. Weigh and count the precipitates.

Standard

A sample of potassium antimonyl tartrate (about 0·5 mg) has been irradiated along with the samples and has been dissolved in 50% HCl,

* The Sb carrier is made by dissolving tartar emetic in water. (Boiling is necessary to complete the solution).

and the volume made up to 10,000 ml. Take 1 ml of the standard solution (equivalent to about 0·02 μg Sb) and carry out the whole procedure, including the digestion step. Counting is carried out with an end window Geiger-Muller counter.

Calculation

The counts of the samples and standards are corrected for decay, counter dead-time, and chemical recovery.

The count rate for the determination is found as follows:

$$\text{Count rate (C.R.)} = \frac{\text{Count standard (c/m)}}{\text{Weight of standard irradiated (mg)}} \times \frac{323{\cdot}93 \times 10^4}{121{\cdot}76}$$

Then the antimony content of the samples is given (in parts per million) by:

$$\text{Sb (p.p.m.)} = \frac{\text{Counts sample (c/m)} \times 10^6}{\text{Weight sample (mg)} \times \text{C.R.}}$$

14

METHOD FOR ARSENIC ANALYSIS

1. Place the activated samples (dried human liver) in digestion flasks.
2. Add 8 ml of digestion mixture (nitric: sulphuric acid, 5:3).
3. Heat till the nitric acid is removed—about $\frac{3}{4}$-hour.
4. Wash out the digested samples into a 200-ml flask, filling the digestion flask three times with water and adding this to the solution in the 200-ml flask.
5. Add 4 ml conc. HCl and 2 ml conc. sulphuric acid.
6. Add 10 μg As carrier ($AsCl_3$) as 1 ml of soln.
7. Add 5 ml 15% sodium iodide, and 0·4 ml 40% stannous chloride in 50% HCl.
8. Wash reagents into flask, the final volume being about 150 ml.
9. Assemble the distillation apparatus, and heat the 200-ml flask in the water bath at its boiling point for 5 min.
10. Add 10 g zinc pellets, insert the stopper and allow reaction to continue for 15 min.
11. Arsenic (in the form of arsine) is removed by passing the gases evolved through a solution of 1·6% mercuric chloride, hydrogen sulphide being absorbed in the lead acetate impregnated cotton wool filter.
12. To the solution in the trap, add 5 ml 0·001 N iodine in 40% sodium iodide.
13. Make the solution up to 20 ml, and count 10 ml.

Standard

A sample (about 0·1 g) of a solution containing 1 mg As per ml has been irradiated along with the samples, and the solution has been made up to 10,000 ml. Take 1 ml of this dilution (i.e. about 0·01 μg As), add 5 ml of the iodine solution as in (12) above, and make up to 20 ml and count 10 ml. The standard was made from $AsCl_3$.

Calculation

The counts of samples and standard are corrected for decay and counter dead time. The count rate (C.R.) for the experiment is determined as follows:

$$\text{C.R.} = \frac{\text{counts (standard) [c/m]}}{\text{weight of As in standard (mg)}}$$

Because the standard is prepared from a solution of 1 mg As per ml which has been diluted by a factor of 10^4, the C.R. is given by:

$$\text{C.R.} = \frac{\text{counts (standard) [c/m]} \times 10^4}{\text{weight standard soln. irradiated (g)}} \text{ cm}^{-1} \text{ mg}^{-1} \text{ As}$$

Then Arsenic concentration (parts per million) in the samples is given by:

$$\text{As (p.p.m.)} = \frac{\text{counts (sample) [c/m]}}{\text{weight sample (mg)}} \times \frac{10^6}{\text{C.R.}}$$

15

METHOD FOR COPPER ANALYSIS

1. Place the samples (dried human liver) in the digestion beakers.
2. Add 10 drops of fuming nitric acid, and evaporate to dryness. Repeat the addition of nitric acid and evaporation, to dissolve the sample completely.
3. Add 10 mg Cu carrier (cupric nitrate) as 1 ml of soln.
4. Add 1 drop of each of the following: 10% cobalt nitrate; 10% manganous nitrate; and 10% ammonium dihydrogen phosphate.
5. Transfer the digested samples with washings to 50-ml centrifuge tubes.
6. Add 3 ml (20%) sodium sulphite solution (if a precipitate forms dissolve it in 1·0 N nitric acid).
7. Add 1 ml (20%) KCNS solution.
8. Boil, spin down the precipitate, and wash it twice with water.
9. Dissolve the precipitate in 0·5 ml conc. nitric acid (added quickly).
10. Add 10 drops of Fe carrier (10% ferric chloride), 2 drops of ammonium dihydrogen phosphate (10%), dilute the solution to make 15–20 ml and add 3 drops of Ca carrier (10% calcium chloride).
11. Neutralize with conc. ammonia solution, adding it slowly at first, and then adding excess.
12. Add 3 ml sodium carbonate solution (10%).
13. Boil the solution on the water bath, spin down the precipitate and filter it off, collecting the supernatant in a clean centrifuge tube.
14. To the solution, add conc. acetic acid to give a pale blue colour.
15. Add 0·5 ml conc. nitric acid and precipitate the cuprous thiocyanate as in (6), (7) and (8) above.
16. Wash the precipitate and dissolve it in 0·5 ml conc. nitric acid.
17. Neutralize to a faint precipitate with 5% sodium hydroxide.
18. Add 4 drops conc. nitric acid and heat.
19. Add 4 ml quinaldic acid solution (2%).
20. Spin down the precipitate and wash it twice with water and once with acetone. Transfer it with acetone to trays, weigh it and count it.

Standard

A sample of copper solution (10 mg Cu per ml) has been irradiated along with the samples and on its return it has been diluted with water and the volume made up to 10,000 ml to give the standard solution. One ml (roughly equivalent to 0·1 μg Cu) of the standard solution is processed starting at step (3). The weight of precipitate recovered from the 10 mg Cu carrier added gives the recovery, and the activity of the samples is determined using an end window Geiger–Muller counter.

Calculation

The counts of the standard and samples are corrected for decay, counter dead-time, and chemical recovery. The count rate for the determination (C.R.) is given by:

$$\text{C.R.} = \frac{\text{Counts standard (c/m)} \times 10^4}{\text{Weight Cu soln. irradiated (g)} \times 10}$$

The factor 10^4 arises due to the dilution of the standard solution and the factor of 10 in the denominator is due to the standard solution containing 10 mg Cu per ml. Then the concentration of copper in the samples in parts per million is given by:

$$\text{Cu (p.p.m.)} = \frac{\text{Counts sample (c/m)} \times 10^6}{\text{Weight sample (mg)} \times \text{C.R.}}$$

16

METHOD FOR GALLIUM ANALYSIS

Samples—Tomato seeds.

1. Preparation of the standard. Break the silica tube with a diamond cutter, and transfer to a small beaker containing 10 ml conc. HNO_3. Boil for 15 min, replenishing the acid if required. Transfer the contents of the beaker to a 1-litre flask, and make up to the mark with water. A 1-ml aliquot is used as standard solution.

2. In a fume cupboard, transfer tissue and standards to 50-ml centrifuge tubes, and boil with 0·5 ml conc. HNO_3. Add 10 mg Ga (0·5 ml Ga carrier solution), and 1 drop each of $Cu(NO_3)_2$, $Mn(NO_3)_2$, $Y(NO_3)_3$ and $NH_4H_2PO_4$, and boil for 5 min. Cool in ice, add an equal volume of 12 N HCl and wash the contents of the tube into a 50-ml separating funnel with 10 ml 6 N HCl.

3. Shake with three 5-ml portions of ether; then run the aqueous layer to waste. Wash the ether layer twice with 5-ml portions of 6 N HCl. Extract the ether layer twice with 5-ml portions of water and pool the aqueous extracts.

4. Boil to expel traces of ether, acidify with HCl, and add 5 drops NH_4HS, 1 drop $Cu(NO_3)_2$, 1 drop ammonium molybdate, and 1 drop formic acid. Spin down CuS, transfer supernatant layer to a fresh tube and boil out H_2S. Add 5 drops $Fe(NO_3)_3$ and 1 drop $NH_4H_2PO_4$, and make alkaline with NaOH. Spin down, add 2 drops $Fe(NO_3)_3$ and re-spin. Pour off supernatant layer into a fresh tube, acidify with HCl, add 2 drops $Fe(NO_3)_3$ and make alkaline with NaOH (solution goes dark brown). Spin down $Fe(OH)_3$ and transfer to a fresh tube.

5. Bring to pH 6 with CH_3COOH/CH_3COONH_4 buffer and add 3 ml oxine solution. Boil and filter through the filter tubes provided. Wash the precipitate with water, transfer to a weighed planchette, dry under the infra-red lamp, and count.

6. Calculate the weight of Ga in the sample using the usual formula, making corrections for dead time and background.

Reagents

Conc. HNO_3	50% w/v $CH_3CO_2NH_4$	10% w/v $NH_4H_2PO_4$
12N HCl	50% w/v NH_4HS	10% w/v ammonium molybdate

6N HCl	50% w/v NaOH	
18N CH_3COOH	10% w/v $Cu(NO_3)_2$	2% w/v 8-hydroxy quinoline in ethanol (oxine)
100% HCOOH	10% w/v $Fe(NO_3)_3$	
Ether	10% w/v $Mn(NO_3)_2$	Ga carrier = 20 mg Ga/ml
	10% w/v $Y(NO_3)_3$	11·47 g $Ga(NO_3)_3 . 8H_2O$ in 100 ml water

17

METHOD FOR GOLD ANALYSIS

Analytical Procedure

About 500 mg of biological ash* were sealed in polyethylene tubes, each of which was covered by 5 mm Al-foil for protection in case the plastic tubing disintegrated during irradiation. The standards were made by pipetting a known volume (10λ–30λ) of gold standard solution prepared from gold wire on to a small piece of filter paper of good quality (15 mm diameter) and by evaporating to dryness. Each filter paper was sealed in plastic and protected in a way similar to the samples. The amount of Au in the standards ranged from 7×10^{-9} g to $2{\cdot}4 \times 10^{-7}$ g.

1. The irradiated ash is to be dissolved by heating in 15 ml of aqua regia, containing 10 mg Au carrier prepared from gold chloride. Dilute to 30 ml with water.

2. Transfer the solution to a separatory funnel and shake for 2 min with 30 ml of ethyl acetate.

3. Wash the organic layer with 15 ml of 6 M HCl. Shake for 1 min.

4. Extract the gold from the organic layer by shaking it with 25 ml of NH_4OH (1:4) for 2 min.

5. Take the aqueous layer and acidify the solution by the addition of 5 ml of conc. HCl.

6. Heat and bubble SO_2 in the solution for 15 min (3–5 bubbles/sec).

7. Filter the deposited Au, wash the precipitate twice with cold 1 M HCl, and mount for counting.

Chemical yield: Weigh as Au metal.

* From human liver.

18

14 MEV NEUTRON ACTIVATION ANALYSIS EXPERIMENTS

Introduction

Activation analysis with the 14 MeV neutron generator, using short-lived radioisotopes and γ-ray spectrometry, forms the basis of a rapid non-destructive method of instrumental analysis. The neutron generators are small accelerators, operating at about 150 kV, in which a titanium tritide target is bombarded with deuterons to produce 14 MeV neutrons by the reaction:

$$^{3}\mathrm{H}(d, \alpha)\,n + \sim 17\ \mathrm{MeV}$$

the remaining 3 MeV of energy being associated with the α-particle. The neutrons are emitted isotropically, and the fast neutron flux at approximately 1 cm from the target is about $\frac{1}{10}$ the total neutron output. Thus, fluxes in the range 10^{9}–10^{10} neutrons cm^{-2} sec^{-1} are common.

Experiment 1

The determination of oxygen is one of the more outstanding of the applications of 14 MeV neutron analysis, because of the extreme simplicity and rapidity of the measurement. The oxygen is measured via the reaction

$$^{16}\mathrm{O}(n, p)\,^{16}\mathrm{N}$$

the nitrogen-16 being detected, following β-decay, by means of the high energy γ-rays (6–7 MeV) of oxygen-16.

(*a*) *Calibration of Spectrometer*. Using a cobalt-60 source, arrange for the 2·5 MeV *sum peak* to fall in channel 80. Decrease amplifier gain four-fold (i.e. channel 80 $\simeq$ 10 MeV).

(*b*) *Irradiation*. Irradiate the rabbit, containing Perspex, for 30 sec and transfer to the detector as quickly as possible, and obtain the γ-ray spectrum, noticing that after about 30 sec the counts no longer build up, as nitrogen-16 has a half-life of only about 7 sec.

Repeat the irradiation, but this time set the analyser at multiscaler (1 channel/sec) and obtain the decay curve and confirm the half-life.

(*c*) *Analysis.* Determine the oxygen content of the unknown sample by irradiating and counting under the same conditions as (*b*). Compare the induced activities:

(i) from the areas of the photopeaks;
(ii) from the sums of all counts above 3·5 MeV;
(iii) from the initial intensities evaluated from the decay curves.

Experiment 2

Barium and phosphorus are common additives in lubricating oil. They can be measured rapidly simultaneously via the reactions

$$^{138}\text{Ba}(n, 2n)\,^{137}\text{Ba}$$

$$^{31}\text{P}(n, \alpha)\,^{28}\text{Al}$$

the induced activities being detected, respectively, via the 0·66 MeV and 1·78 MeV γ-rays.

(*a*) *Calibration.* Use caesium-137 (0·66 MeV) and sodium-22 (0·51, 1·28 and 1·79 MeV) for calibration.

(*b*) *Irradiation.* Irradiate the rabbit, containing the barium standard, for 1·0 min, allow 30 sec for shorter-lived activities to decay and then count for 1·0 min, observing the 0·66 MeV photopeak.

Repeat using the phosphorus standard but observing the 1·78 MeV photopeak.

(*c*) *Analysis.*

(i) Determine the barium and phosphorus contents by comparison of the photopeak areas of the full spectrum with those of the standard.
(ii) Measure the two half-lives by selective counting of the 0·66 MeV and 1·78 MeV photopeaks.

19

THE DETERMINATION OF TRACES OF COPPER BY ISOTOPIC DILUTION ANALYSIS (SUBSTOICHIOMETRIC METHOD)

Introduction

To analyse a sample for a particular element a known amount of radioactive labelled element is added in such a form that complete isotopic exchange will occur. After complete mixing, the element is separated and its specific activity determined. Since the total activity is unchanged by the mixing, the specific activities can be related as follows:

$$\text{Total activity} = S_1 \times X = S_2 \times (X+W)$$

where S_1 = specific activity of added material
S_2 = specific activity of mixture
X = weight of element added
W = weight of element in unknown

whence

$$W = X\left(\frac{S_1}{S_2}-1\right)$$

The principle of the substoichiometric method is that an insufficiency of reagent is used to separate equal amounts of the element from the original tracer solution and from the mixture. By this means the necessity of determining chemical yields of separations is avoided and the measured counting rates of the samples can be substituted in the above equation in place of specific activities. As an example of this technique traces of copper are determined by a solvent extraction procedure in which an insufficiency of extracting reagent (dithizone) is used.

Experimental

Reagents, demineralized water, carbon tetrachloride, A.R., dithizone, A.R.

Buffer solution: dissolve 15 g of potassium iodide, A.R., 13 g of ammonium acetate, A.R., in 100 ml of water, and add glacial acetic acid,

A.R., to attain a pH value of $4{\cdot}8 + 0{\cdot}2$. This solution should be purified before commencing the experiment by extraction with dithizone.

Radioactive copper solution: dissolve irradiated copper sulphate in acidified water and dilute to the required concentration.

Non-active copper solution: dissolve copper sulphate in acidified water and dilute to the required concentration.

The "unknown" copper solution to be analysed should be made approximately 10^{-7} g of copper per ml. A solution of radioactive copper of stated concentration will be provided, and if necessary should be diluted to give a solution of the order of 10^{-6} g copper per ml. A solution of 5×10^{-4} M dithizone will be provided. A portion of this should be diluted with carbon tetrachloride to give a 2×10^{-5} M solution of dithizone. The buffer solution should be purified by extractions with the concentrated dithizone solution until the extract is not discoloured.

Procedure

1. Setting up counting equipment

Place 1·0 ml of the radioactive copper solution in a suitable container, and use this to determine the optimum conditions for counting in the scintillation counter provided.

2. Analysis of "unknown" solution

Transfer 10 ml of the test solution to a separatory funnel and add 1·0 ml of the radioactive copper solution. Add 1·0 ml of buffer solution and shake for 5 min with 1·0 ml of 2×10^{-5} M dithizone solution. Take 0·5 ml of the organic phase and determine its radioactivity content (S_2).

3. Counting of tracer solution

Take 1·0 ml of the radioactive copper solution and dilute to 10 ml with demineralized water. Proceed as above to determine its radioactivity content (S_1).

4. Results

Calculate the copper content of the "unknown" solution by the equation given above, and comment on the result.

Notes

The specific activity of the radioactive copper solution provided will be just sufficient to determine the amount of copper present in the unknown. Higher specific activity copper can be obtained to determine even smaller concentrations of copper.

FURTHER READING

1. Ruzicka, J. and Stary, J. (1962). "Isotope dilution analysis by solvent extraction—IV. Selective determination of traces of copper with dithizone," *Talanta*, **9**, 617.

20

INVESTIGATION OF SOME PRODUCTS FROM RADIOACTIVE DECAY OF NATURAL THORIUM

Apparatus

30 ml platinum crucible, platinum foil electrode, rheostat, meter and three 2-V accumulators. End window Geiger counter. Scaler.

Object

The object of the experiment is to separate the lead and bismuth activities, and to characterize them by observing their decay. See Fig. 46.

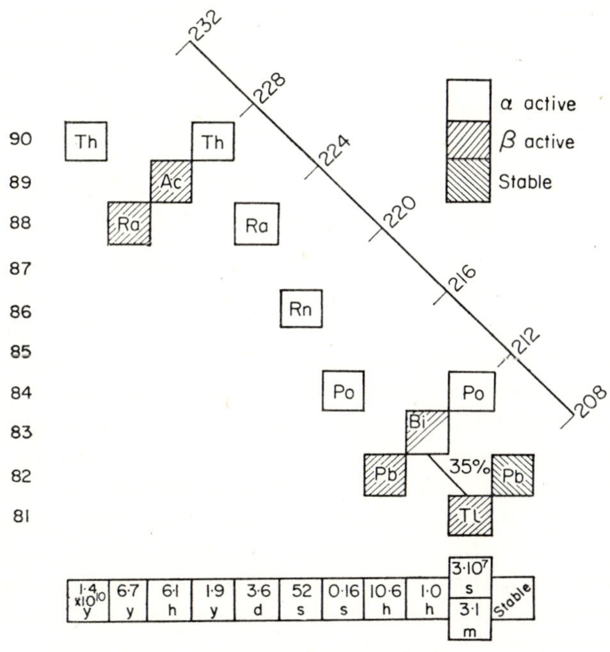

Fig. 46. Thorium-232 radioactive decay scheme.

Procedure

About 20 ml of the thorium solution provided (0·5 M thorium nitrate in 0·1 N nitric acid) is placed in the platinum crucible which serves as the cathode in electrolysis. The anode consists of a platinum foil electrode. Current is supplied by three 2-V accumulators in series and adjusted, with the rheostat, to 0·1 amp. The solution is electrolysed for 10 min, then the anode is removed and rinsed into the wide-mouth bottle labelled "Thorium Rinsings". The solution in the crucible is poured into the bottle labelled "Electrolysed Thorium Nitrate". The crucible is rinsed several times with distilled water, and the rinsings added to the "Thorium Rinsings" bottle. The anode is placed in a narrow test tube containing sufficient sodium nitrite solution (8 gm/litre) to cover that part of the electrode which was immersed during electrolysis. About 4 drops of conc. nitric acid are added and the solution stirred (with the electrode) after the addition of each drop. The electrode is allowed to stand in this nitrous acid solution for several minutes.

One ml of conc. nitric acid is added to the crucible, heated just to boiling and swirled in the crucible for a few minutes to dissolve completely the tracer amounts of lead and bismuth activity. During this procedure the crucible should be held in the steel-wire ring provided, rather than with tongs, to minimize the risk of dropping the crucible and spilling the contents. The nitrous acid in which the anode was washed is then added to the crucible.

Use of Carrier Techniques to Separate Lead and Bismuth Activities

One ml each of 0·1 N lead nitrate and 0·1 N bismuth nitrate is added to the nitric acid solution in the crucible. Six drops of conc. sulphuric acid are added and the mixture is heated until fumes of SO_3 are given off. Care is necessary to avoid spattering. After allowing to cool, 1 ml water is added and the mixture transferred to a centrifuge tube, washing in all of the precipitate with 0·1 M sulphuric acid. After centrifuging, the supernatant liquid, which contains the bismuth, is drawn off with a semimicro pipette, transferred to another centrifuge tube and saved.

The precipitate of lead sulphate is washed twice with 0·1 M sulphuric acid, using the centrifuge to separate off the washings which are discarded. The lead sulphate is made into a slurry with about 0·5 ml acetone, and transferred, using a semimicro pipette, to a counting planchette and evaporated to dryness under a heat lamp.

To the centrifuge tube containing the bismuth activity is added conc.

ammonia solution until the solution is alkaline. The mixture is centrifuged and the supernatant liquid discarded. The precipitate of bismuth hydroxide is washed three times with distilled water and transferred as an acetone slurry to a planchette and dried under the heat lamp.

Counting of the Sources

1. Each of the planchettes is mounted on a Perspex holder, and when the sources are changed, as described below, steps are taken to ensure that they are subsequently replaced in exactly the same position so that the counting geometry will be standard.

2. Beginning at an arbitrary zero time, the bismuth source is counted for 10 min and the count rate noted. During the next two minutes the bismuth source is replaced by the lead source, and beginning at 12 min on the arbitrary scale, this source is counted for 10 min. The procedure is repeated for the bismuth source after a further two minutes, and so on. By this means, activities are determined with respect to time in units of fifths of hours. After three hours the interval between successive counts may be increased, and it may be possible to obtain an accurate estimate of the half-life of the bismuth without counting this source any further. However, at least one count rate determination should be made the following day on the lead source.

3. A background count rate is determined, and the results of (2), corrected for background and dead-time, are plotted against time on semilogarithmic paper. (Count rate on log. axis.)

4. Determine values for the half-life of each isotope and observe any peculiarities in the shapes of the graphs.

21

MEASUREMENT OF THE HALF-LIFE OF PROTACTINIUM-234

Apparatus

Cation-exchange column, containing Amberlite IR-120 resin (a sulphonated polystyrene). Liquid flow counter. Scaler.

Theory

^{234}Pa is one of the decay products of ^{238}U (the most abundant isotope in natural uranium). The decay scheme Fig. 47 is:

$$^{238}\text{U} \xrightarrow[t_{1/2}=4\cdot5\times10^{9}y]{\alpha} {}^{234}\text{Th} \xrightarrow[t_{1/2}=24d]{\beta} {}^{234}\text{Pa} \xrightarrow[t_{1/2}=?\text{(short)}]{\beta} {}^{234}\text{U} \xrightarrow[t_{1/2}=2\cdot5\times10^{5}y]{\alpha} {}^{230}\text{Th}$$

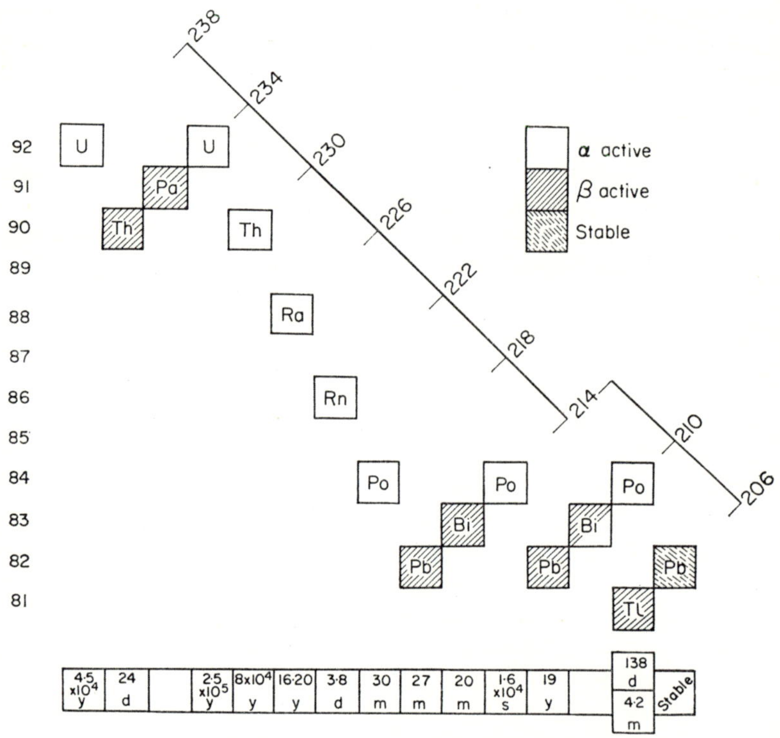

Fig. 47. Uranium-238 radioactive decay scheme.

followed by ten more stages to ^{206}Pb (stable). In refined uranium the series effectively ends at ^{234}U because of the long half-life of this isotope.

Object

The object of the experiment is to separate ^{234}Pa from its parent ^{234}Th, and to observe its rate of decay.

Procedure

PART A

1. The resin in the column is washed with 200 ml distilled water.
2. 100 ml of a solution containing 20 g uranyl nitrate is allowed to flow through the column at a rate of 2–3 drops per sec.
3. The effluent from the column at this *and each subsequent stage* is poured into the "uranium residues" bottle.
4. The column is washed successively with 100 ml of water, 200 ml of 0·3 N sulphuric acid and 100 ml water at a rate of about 5 drops per sec. (This elutes the uranyl ion leaving thorium on the column.)

Since the half-life of ^{234}Th is 24·1 days, the prepared column can be used as a source of ^{234}Pa for several weeks, without adding any more uranium solution.

PART B

1. Measure the background count rate using the liquid-flow counter.
2. Fill the reservoir with 5 N hydrochloric acid and start the scaler.
3. Admit HCl to the column, with the tap open so that the effluent from the column passes through the counter. This will elute ^{234}Pa.
4. Observe the count rate qualitatively until a reasonable activity is indicated. (4–5 min after first admitting HCl).
5. Close the tap to stop the flow through the counter and begin counting immediately.
6. Count for 15 sec, read and reset the scaler in the next 5 sec, and so on, so as to obtain a series of 15 sec counts at 20 sec intervals until the count rate has fallen to the background value.
7. Since the protactinium is short-lived, it quickly grows back to its equilibrium activity in the column. The experiment can therefore be repeated after 10 min if necessary, simply by opening the tap for a few minutes to admit a fresh sample to the counter. The results of the two experiments may be plotted on the same graph.
8. Pour all effluent into the "uranium residues" bottle.
9. Plot the activity vs. time on semilogarithmic paper in the usual way, and determine the half-life.

INDEX